A Treatise on Human Life

An Unalienable Right

By Harold D. Kletschka, M.D.

Alethos Press LLC

ALETHOS PRESS LLC
PO Box 600160
St. Paul, MN 55106
http://www.alethospress.com

A TREATISE ON HUMAN LIFE - AN UNALIENABLE RIGHT

ISBN 0-9702509-2-4, Hardcover
ISBN 0-9702509-3-2, Paperback

http://www.treatiseonlife.com

Printed in the United States of America

10 9 8 7 6 5 4 3 2 1

This Work is Respectfully Dedicated to

GOD

The Creator of All Life
That all Life may be Protected and Preserved

and

In Awesome Gratitude for:

All the lives He has brought and will bring into existence

The life He gave me

The parents he gave me;
my deceased Father, Herbert and my Mother, Emma
who taught and showed me the value of life

The sisters He gave me,
Virginia, Marjorie, and Barbara
who enriched my life

A Treatise on Human Life

CONTENTS

SECTION III

An Analysis of Roe v. Wade

Introduction

The right man for this treatise
By Dave Racer

"We hold these truths to be self-evident, that all men are created equal, that they are endowed by their Creator with certain unalienable Rights, that among these are Life, Liberty and the pursuit of Happiness..."

The American Declaration of Independence, 1776

America's founding fathers were, for the most part, men of Christian faith. They were serious men of medicine, history, science, philosophy, and the law.

Among the signers of the Declaration were: Benjamin Rush, a medical doctor; Benjamin Franklin, who became known as much for his scientific experiments and inventions as his wide range of political and governmental achievements; James Wilson, a classical scholar and teacher before becoming a lawyer who became one of the original members of the United States Supreme Court; John Witherspoon, who served as the President of Princeton University, but by training was a pastor. These four founding fathers, then, represented medicine, science, law and theology.

From the eclectic mix of talents and training of these four and the other 52 signers came the marvelous Declaration, filled as it was with philosophy, theology, law, politics and diplomacy. As a result, a new nation was born that was "conceived in liberty and dedicated to the proposition that all men are created equal," as Abraham Lincoln so succinctly stated it.

To the founding fathers, the critical issues about human life had long-since been settled, whether as part of "the laws of nature and of Nature's God," Biblical teachings, or in the Common Law which they adopted from Great Britain. Because of their general agreement and understanding about these issues, they were able to write a federal constitution of enumerated rights that laid out general boundaries for the new government, rather than create a lengthy document that intended to specify every aspect of government authority. Their acceptance of self-evident truth — and the Common Law — made this generalized constitution possible.

From the beginning, politics and man's natural tendency toward power, greed and self-indulgence attempted to erode the new constitution. Yet, it stood firmly against man's worst proclivities because foundational to it all, American citizens knew they could always appeal to the law, a law that had been built on immutable principles and preserved so well for more than a millennium in the Common Law. Yet, fealty to the Common Law also eroded during the Nineteenth Century as federal judges began to assert their own human philosophies into interpretation of legal matters.

When the U.S. Supreme Court handed down its 1973 Roe v. Wade decision it wiped clean multiple centuries of legal precedent. From universal condemnation of abortion to the Court's sanctioning of legal abortions, the Court had rendered *stare decisis* moot. Instead, by the whim of seven justices, the foolishness of interest-group politics was substituted for the wisdom of the ages. As tragic is the loss of millions of human lives through abortion, the Court's substitution of man's will for immutable law has wreaked an even more far-ranging devastation on freedom and liberty; it has rendered futile a predictable future.

Into this legal void stepped a man of conviction, one who combined the skills, talents and training of the founding fathers. Twenty years ago, Harold D. Kletschka, M.D., a diminutive and unassuming Minnesotan who loves truth, applied his well-honed talents to finding the truth about the Law and human life.

Dr. Kletschka is a medical doctor, a 1947 graduate of the University of Minnesota Medical School, who went on to become a board-certified general, thoracic and cardiac surgeon. This, then, is a trait he shares with Benjamin Rush.

In 1957, Dr. Kletschka conceived of the world's most perfect artificial heart. Along with bio-medical engineer Edson Rafferty, he conducted hundreds of experiments on pulmonary and cardio-vascular systems, to test and prove the efficacy of his artificial heart. By 1975, he and Rafferty had proven the superior efficiency of the Bio-Pump®, their heart assist device that today is used in nearly half of all open-heart surgeries worldwide. He has authored and had published numerous medical journal articles reporting on his heart and blood experiments. During the 1990s, he perfected a new angioplasty device which even today is being readied for clinical testing. As a research scientist, then, Dr. Kletschka takes his place alongside Benjamin Franklin. (Kletschka, like Franklin, is a man of immense curiosity about nearly every aspect of humanity — politics, government, theology, science, history, literature, and writing).

In 1970, Blackstone School of Law bestowed a Bachelor of Law Degree on Dr. Kletschka. The Blackstone school took its name from Sir William Blackstone, whose 1758 lectures at Oxford were later published as the "Commentaries on the Laws of England." His "Commentaries" and legal dictionary encompassing the extant English law, provided an important foundation for American jurisprudence, and he is considered the foremost historical expert on the Common Law. Although the Blackstone School of Law was not accredited for taking the bar, it provided Dr. Kletschka with a reliable and indispensable legal background in preparation for his career as Chairman of the Board, President and Chief Executive Officer of Bio-Medicus, Inc., an international company devoted to the development and marketing of medical products. By training, then, in 1970 Dr. Kletschka joined James Wilson as a man of the law.

(Interestingly, Dr. Kletschka's enormous drive to learn and master subjects did not limit him to medicine, science and the law. In 1973, the United States Air Force named Air Force Reserve Colonel Kletschka as the worldwide outstanding graduate of his Air War College class. His contributions and achievements also resulted in his inclusion in Marquis' "Who's Who in the World.")

Throughout most of his life, Dr. Kletschka has been a devout Roman Catholic. Following his father's death in 1960, Dr. Kletschka consecrated his work on the artificial heart to Our Lady of Mount Carmel, the particular title under which Harold had a special devotion to The Blessed Virgin Mary. While he would never claim the title "theologian," Dr. Kletschka nonetheless has devoted thousands of hours to the study of his faith and religion, and is an effective apologist for his faith on critical issues. Like John Witherspoon, then, he is a man of theology.

In Dr. Kletschka, then, we find a unique combination of physician, scientist, lawyer and theologian. While passionate about protecting human life and preserving the integrity of the law, seeing life through the eyes of faith, his training as a scientist impels him to test all things. That is why, following the 1973 Roe v. Wade decision, and after years of contemplation, he set about to do the arduous work reflected in this, "A Treatise on Human Life." For his study he chose primary documents, ancient sources dating back to the Fourth Century and earlier. He sought the best counsel of learned men and women to assist him in the translation of these sometimes-difficult texts. He set aside his personal bias and assumptions to examine facts that could be tested from multiple angles.

Dr. Kletschka's 20-year quest for the truth about human life in the womb, Common Law, and the American system of justice has convinced him, and ought to convince any serious and honest reader, that Roe v. Wade was an attack on liberty that needs to be set right, that is, if America is to remain a nation of free people with leaders who are bound down by law. Because, as Rush, Franklin, Wilson and Witherspoon all knew, the self-evident truth of the unique sanctity of human life is the starting point of liberty.

—

Dave Racer is an author, commentator and political consultant from St. Paul, Minnesota. Among his many writings is the biography of Harold D. Kletschka, M.D., To Change the Heart of Man, *to be published during 2003.*

Preface

This treatise is designed to be a definitive work on the legal status of abortion in America based on the common law it adopted from England and the necessary relationship it bears to the unalienable right to life of a human being. To achieve this ambitious goal a comprehensive research effort was undertaken including a diligent study of the development of the common law governing abortion from the legal, historical, ecclesiastical, Biblical, religious, biological (and medical), political, and cultural standpoints and the bearing each had in the formulation of the precedents established. Utilizing the information gathered from each of those sources, the actual common law precedents as they were originally established and then perpetuated are documented, supplemented by comments to explain significant relationships, clarify points, or to emphasize some aspect of the law. Only prime sources of enactments, decrees, or declarations were held to be acceptable. This requirement eliminated bias based on private interpretation.

The earliest laws required translation of Latin from the original Anglo-Saxon language. Once the earliest common law rulings on abortion were identified, succeeding rulings down through the ages were catalogued and recorded. This chain of common law rulings governing abortion were traced from their inception, at the earliest days of the Anglo-Saxon society, up to the present time, and it was found that abortion under the common law has been a crime, without break in continuity, up to the most recent times, although often courts did not fulfill their obligation to adhere to the established law, contributing, in part, to confusion among various legal scholars as to what constituted the binding common law. Their confusion would have been erased had they simply gone back to the original common law and traced its history, century by century.

Recent legal scholars have universally been deceived by marking the beginning of common law on abortion to the medieval ages when secular courts assumed jurisdiction of those cases, which had previously been handled exclusively by the ecclesiastical courts. The secular courts, however, were bound to follow precisely the precedents already established by the ecclesiastical courts and ancient kings. However, they did modify the punishments to be meted out for the crime of abortion in an attempt to secure more convictions. Recent scholars have interpreted the medieval courts as considering the crime of abortion in a less serious light based on this reduced punishment, whereas, in actuality, it represented a hardening of the position taken by the medieval courts, taking a more serious view of the crime reflected in its attempt to obtain a greater number of convictions for the offense.

In 1970, it first came to the attention of the author of initial steps taken to disassemble the abortion laws, learning that Hawaii began allowing "legal" abortions, and that several other states had revised their statutes to decriminalize the act. It did not seem that those deviations from long accepted practices over many centuries would be long lived. The author felt these aberrational views were simply the result of the rebellious and regressive spirit of the "me" generation that infected America in the 1960's when problems were all to be solved by slogans and clichés, and the right of freedom was interpreted as a license to satisfy every personal desire without restraint. Having been reared in a society and environment in which abortion was viewed as a heinous act, it seemed certain these new rulings would be overturned on appeal to the U. S. Supreme Court where these innovations would come under review by learned and detached legal scholars.

Needless to say, in 1973, when the Roe v. Wade decision was issued, legalizing abortion, it came as a profound shock to the author, as he realized this august tribunal had been cowed by the tidal wave of superficial emotionalism then sweeping the country. It did not seem realistic that even that decision would hold for long because it was so contrary to common sense, and the legal and moral condemnations of abortion, which had extended over a period of almost two millennia in the Western world. However, surprisingly, it was not only upheld in follow-on decisions, but permissiveness to commit the act was actually extended, so that it became firmly rooted in the judicial system, and much of our general society seemed to have accommodated their views to coincide with those accepting abortion as an acceptable procedure, supported by what they felt was the weight of legal precedent.

Pro-life voices emerged, and their arguments condemning abortion were universally weighty and, seemingly, to a rational mind, sufficient to dispel the myth that abortion was a right under the constitutional guarantees as ruled by the Supreme Court. However, the pro-abortion voices, relying on slogans, and pandering to a society which had become educated and acclimated to accept shallow and unsupportable claims and propaganda in order to excuse its incessant insistence of being able to satisfy its sensual, and even carnal, pleasures without carrying a price tag of self-sacrifice or obligation to another human being, were able to extend their influence to mute the voices of opposition.

The pro-life groups propounded very significant arguments in support of their position, but generally these efforts appeared to the author as a patchwork of labors. Some condemned the practice of abortion from a purely religious perspective. Others found their basis for opposition in constitutional law. Others relied on decisions from older legal precedents, e.g., the statements found in Blackstone's *Commentaries*. Still others cited medical or biological findings in support of their position. Very distressing were the concessions so often unwittingly made by those condemning abortions, because the stand they would so often concede actually amounted to forfeiture of a position historically in accordance with their own. Too many of the pro-life arguments, it seemed to the author, were simply emotional emanations of justifiable feelings, although not supported from a compelling factual standpoint.

Because of the failure of segmented, patchwork, emotional, and often irrelevant arguments to set forth the controlling principles governing abortion, it prompted the author to look further into the matter. It seemed inconceivable to him that the laws unremittingly condemning abortion through almost two millennia, and even, in practice, being condemned in pagan and uncivilized countries dating back to the pre-Christian era could be suddenly overturned on some newly discovered reason, especially since legal precedent has to control in the American judicial system, and advances in the biological sciences actually added support to the pro-life position.

Therefore, about twenty years ago, the author undertook an all-encompassing investigation covering all pertinent facts bearing on the subject as earlier discussed. The review has included search into many sources beyond those cited herein. There was never found anything in the excluded reference works that could successfully contradict or impugn the credibility of the established common law materials on abortion that were selected for inclusion in this treatise. Not surprisingly, the selected referenced works accurately and indisputably show that the actual common law governing abortion in America is permanently established as a crime, Roe v. Wade notwithstanding.

The treatise is divided into Sections I, II, and III.

Section I is set out simply to define human life, human being, and personhood as a separate topic. Being principally based on medical facts it has a stand-alone value, but, once developed, the information shows a magnificent compatibility with the facts pertinent to the common law governing abortion, while no less defining the sacredness of human life. Biblical sources are quoted and analyzed to treat of personhood at the theological and spiritual levels as apart from a purely legal standpoint. It seemed well to give some attention to this aspect of the subject, so as to show how these depictions tied into the essential controlling principles which formed the underpinnings on which America was founded, while always remembering, the common law on abortion was deeply rooted in religious, spiritual, and Biblical values. It seemed to the author this would have a value in and of itself in bringing into perspective, and gaining an understanding of, the sacredness of human life apart from the purely common law history.

Section II is devoted to tracing the unbroken chain of precedents of common law governing abortion dating back to the earliest recorded history on the subject, how this law and precedents were adopted at the time of the foundation of our country, and continue in force today, Roe v. Wade, notwithstanding. Commentaries are used generously by the author to highlight points. Necessarily, there is an overlap of discussion covered in the commentaries in Sections I, II, and III because the information in any given Section often has application in the material covered in the other Sections. Repetition of statements or points is made throughout the text. This exercise has been adopted to assist in maintaining continuity of thought when dealing with a subject without having to simultaneously or repetitively sort through pages of the document as would be necessary if repeated references had to be continuously tracked. At times, the repetitiveness is used to emphasize a point. **What may seemingly be annoying repetitiveness of corrective comments in material covered is necessary because of the repeated restating of errors by the U. S. Supreme Court and authors of the articles being analyzed.** Finally, a summarization of the material covered is made, and conclusions together with recommendations are contributed at the conclusion of Section II, which are compatible with the controlling common laws governing abortion.

Section III consists of an analysis of Roe v. Wade, and demonstrates the invalidity of the decision based on the legal arguments made by the U. S. Supreme Court. Conclusions and recommendations are then added which harmonize with the existing common law governing abortion. It is repeatedly emphasized that from the earliest times in the development of the common law, and reaffirmed on many occasions (E.g., Magna Charta, Coke's *Institutes*, Blackstone's *Commentaries*, etc.), explicit rulings have repeatedly been handed down guaranteeing the perpetual force of the established common law which thereby assures that abortion is a crime at any stage of gestation, that contraception is homicide, and that the father has rights attached to the conceptus. The irrelevancy of Privacy, invented as a "reason" for finding federal jurisdiction, is shown. Proof is set out demonstrating conclusively that life and personhood both exist in the conceptus from the moment of conception (or fertilization).

Throughout this treatise the term, conceptus, is used and is meant to cover all periods of gestation from fertilization through the final stage of pregnancy. This option has been elected to simplify the discussion of the subject matter, and, if a precise time, in a given instance, is intended to be identified by the term, the specific time period can be appreciated in the context of the discussion surrounding the topic at hand.

Harold D. Kletschka, M. D.

Minneapolis, Minnesota
February 10, 2003

Acknowledgments

I want to thank my sister, Barbara, for the devoted time she has spent reviewing and critiquing this work, and the many helpful and indispensable suggestions she has made. My Mother and Barbara have both sacrificed all diversions for many months while bearing with me in the preparation of this manuscript. Their thoughtfulness and encouragement have been a source of inestimable help.

David Racer, a scholar and man of high principle, has lent a sympathetic ear and been a valuable sounding board for assessing the significance of subject matter of this treatise. He first triggered the thought concerning the significance of the predominately non-Catholic Founding Fathers of America adopting the common law of England which had its origin based on Roman Catholic doctrines and teachings.

Special thanks are given to Father Theodore Campbell for his initial review of some of the Latin excerpts and to suggest sources for definitive translations of the texts.

I am indebted to Dr. Kenneth Pennington, Professor of Ecclesiastical and Legal History, Columbus School of Law, at the Catholic University of America, for the help and enlightenment he provided on the Canon Law of the Catholic Church, Papal Decretals, the Gratian Decretum, and history of Medieval Canon Law. Thanks to Bruce Miller, Librarian at the Catholic University of America, for supplying Papal documents and valuable insights on Medieval Canon Law.

Definitive translations of the Latin texts were carried out by Dr. Mary C. Preus, teacher of Latin and Greek studies at the University of St. Thomas, St. Paul, Minnesota. Because of the ancient sources from which the Latin texts were extracted, this undertaking provided formidable challenges, and it was a pleasure working with her on the project as she brought to light the hidden teachings contained in them.

Appreciation is also owed to Sister Mary Forman, Assistant Professor of Monastic Studies at St. John's University, Collegeville,

Minnesota for her translation of certain medieval Latin texts.

Sincere thanks go to the staff at the University of Minnesota Law Library for their help. Suzanne Thorpe, Associate Director; George Jackson, Reference Librarian; Mary Rumsey, Foreign International and Comparative Law Librarian; and Katherine Hedin, Curator of Rare Books supported with enthusiasm my library searches for books and reference materials sparing me countless hours of labor. The zeal with which Ms. Hedin uncovered and made available ancient legal manuscripts was outstanding.

Section I

The Definitive Identification of Human Life, Human Being, and Person

A HUMAN LIFE = A HUMAN BEING = A PERSON

The basic unit of *all* life is the cell. Human cells are different from plant or animal cells. At conception, unique cells—a human sperm and a human ovum—unite to become a human cell and, therefore, being the unit of life, a priori, a human life. Even primitive and ancient societies, before the discovery of the cell, recognized the distinction between vegetative life, animal life, and human life. A sperm, a unicell, has life, but does not possess the genetic code of Homo sapiens, and, although, it may potentially lead to human life, it is not existent human life. The ovum, a unicell, has life but does not have the genetic code of Homo sapiens, and, although, may potentially lead to human life, it is not existent human life. When the sperm and ovum unite in conception a zygote—a unicell—is formed. This unicell has life, but it also carries the Homo sapiens genetic code. Thus, it is human life. The individual human being begins with the formation of the zygote. In the case of identical twins, each individual begins at the time each zygote fails to divide into further zygotes; each zygote results in distinct human beings for each new zygote formed, marked from the time that further zygotes cease to be created by further division.

That is, the zygote, being human life, defines a human being. *HUMAN LIFE RESULTING FROM THE FORMATION OF A ZYGOTE DEFINES THE PARTICULAR DISTINCTION IDENTIFI-*

ABLE AS A HUMAN BEING. A HUMAN PERSON IS A CONTINUUM OF HUMAN LIFE (OF A HUMAN BEING) FROM THE TIME OF FORMATION OF THE ZYGOTE, AND PROCEEDING UNENDINGLY THEREAFTER (IN LEGAL PRACTICE A FICTION HAS BEEN ADOPTED THAT A PERSON CEASES TO EXIST AT THE TIME OF PHYSICAL DEATH). Life is a state of continual change. Inasmuch as only a given human life, i.e., a human being, can be a person, it follows that a person is fully such from conception, even though at that point the person's form lacks all of its potential physical and mental attributes—and nothing, in perpetuity, can change these facts.

Each individual's full potential and distinctiveness is defined at the moment of the completed formation of the zygote. This equates with the completion of the conception process. In the formation of a particular individual, conception is ended with the completed formation of the zygote. But when multiple identical zygotes are formed, from a single sperm and a single ovum, the conception process is completed only when a given zygote no longer undergoes further division. Even though after conception (or fertilization) multiple identical zygotes may be formed out of the union of a single sperm and a single ovum, each zygote has a distinct identity—that is, each, immediately, becomes a distinct person (and this fact is known by common knowledge that one person does not simultaneously co-exist in separate bodies). Each develops into a separate and recognizable person even though they may hold an identical genetic code. This fact demonstrates that being a person—a living human being—has a spiritual dimension inasmuch as the two (or more) identical physical genetic codes results in two (or more) separate cognizable individuals. The person's life progresses as a continuum of manifest changes, yet being the same person all the time. At birth, the human life is a helpless baby, but no less a person than later when he/she can creep, or later still when he/she can walk, or even later when he/she can procreate, or finally when he/she has gray hair or enters a decrepit state of old age. Therefore, it is spurious reasoning to arbitrarily assign nonpersonhood to human life, in utero, which is still proceeding in its ordinary continuum of development, which may be expressed symbolically as follows:

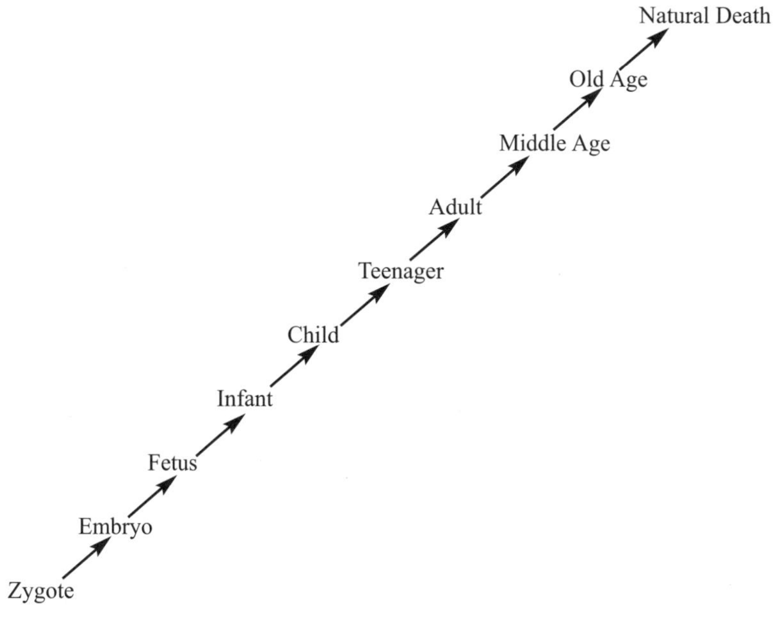

At a certain point in a person's natural development a self-identity ("I am") awareness occurs for each individual. But Rene DesCartes' observation of man: *Cognito, ergo sum,* "I think, therefore I am" is erroneous. A person "is" even without being cognizant of it (E.g., head injury, coma, disease affecting the brain, or early phases of one's existence at which time the person is unaware of his/her existence.). "Thinking" is not a prerequisite for personhood. True enough, a human person's eventual particular ability to think sets him apart from other life forms, but there is a far more important reality to consider.

One can borrow the "I am" (lower case) concept to put forth the idea that at the point of conception, a human becomes "I am". It is compatible with the thesis that human persons never die, but that the form of their life simply changes at the moment of earthly death. The "I am" (lower case) characterizing one's human identity, can be thought of as bearing a relationship to "I AM", the name God called Himself when speaking to Moses. God created man in His Own

Image and Likeness, and the identity which comprises each "I am" perhaps reflects (constitutes?) the image and likeness of "I AM". With or without awareness, this self-identity exists and encompasses the physical, as well as a spiritual quality.

"I AM" is the name God called Himself, and God created man—as little "I am's"—in His image, in essence as complete beings in perfect non-Divine fullness.

It is interesting to note that in an earlier period when objectivity, reasoning based on factual information, and intellectual honesty dominated the composition of the intellectual community, the statement of, Dr. Arey, a world renowned embryologist, found in the standard textbook we used in medical school commands attention. He therein relates: "The formation, maturation and meeting of the male and female germ cells are all preliminary to their actual union into a *zygote* which definitely marks the beginning of a new individual."[1] (Emphasis in original). In accordance with what has already been previously pointed out, he also states, "Although the most striking changes in the development of man and mammals occur while the young (first called an *embryo* and later a *fetus*) is still inside its mother's womb, yet development by no means ceases at birth. **Birth is a mere incident in the whole developmental program, which occurs when the new individual is sufficiently advanced to cope with the conditions of life in the external world.**"[2] (Emphasis added by bold letters). This textbook was published in 1942, long before the unscholarly, polluted, and aberrational reasoning processes infected our courts and culture.

Professor Jerome Lejeune, M.D., Ph.D., discovered the first chromosomal error related to a disease, Down's Syndrome. He was Professor of Fundamental Genetics at the University of Paris Medical School, an internationally renowned authority in genetics, and a pioneer in the field. He testified under oath as to the medical view of whether an embryo is a human being.[3] His testimony on August 10,

[1]Arey, L. B., Developmental Anatomy (Philadelphia: W. B. Saunders Company, 1942), p. 43.
[2]Ibid., p. 1.
[3]Lejeune, J., Testimony in Davis v. Davis, et al., Cir. Ct. Blount Co., Tenn., Equity Div. (Div. I), No. E. 14496, 1989.

1989, is best appreciated by reading his actual statements from the trial court record:

In an early part of his testimony:

> Q.: Dr. Lejeune, let me make sure I understand what you are telling us, that the zygote should be treated with the same respect as an adult human being?
> A.: I'm not telling you that because I'm not in a position of knowing that. **I'm telling you, he is a human being**, and then it is a Justice who will tell whether this human being has the same rights as the others. (Emphasis added). If you make difference between human beings, that is, on your own to prove the reasons why you make that difference. **But as a geneticist you ask me whether this human being is a human, and I would tell you that because he is a being and being human, he is a human being**. (Emphasis added).

Then, in a later part of his testimony:

> Q.: I take it, Dr. Lejeune, therefore, if you believed that a embryo was not a human being as that term is used in ethical or legal or moral or philosophical or religious way that your view of this case may well be different?
> A.: Totally. If I was convinced that those early human beings are, in fact, piece of properties, well, property can be discarded, there is no interest for me as a geneticist. But if they are human beings, what they are, then they cannot be considered as property. They need custody.
> Q.: What it really turns on is what philosophically, ethically, legally that embryo may be. **In your mind, sir, you have come to the very firm conviction that the early embryo or that the embryo is a human being, early human being, as you described it?** (Emphasis added).
> A.: **Yes.** (Emphasis added).
> Q.: And you do recognize in other men's minds, after long and deep thought, learned men, they come to the opposite conclusion you do?
> A.: **No, I don't agree with that.** (Emphasis added).
> Q.: You don't agree with that?
> A.: **I have not yet seen any scientist coming to the opinion**

> **that it is a property**. (Emphasis added). It is what is the case. It's whether they are property that can be discarded, or whether they're human being who must be given to custody. That is it. You ask my question, I answer precisely; **I have never heard one of my colleagues—we differ on opinion of many things, but I have never heard one of them telling me or telling to any other that a frozen embryo was the property of somebody**, that it could be sold, that it could be destroyed like a property, never. (Emphasis added). I never heard it.
> Q.: Just so I understand what you're telling us, I take it, Dr. Lejeune, from your testimony that you would be opposed to abortion?
> A.: Oh, I dislike to kill anybody. That is very true, sir.
> Q.: You would bclicvc that abortion should not be legal?
> A.: That is another point which is different. I think abortion is killing people, and I think in a good jurisdiction would make those killing people become rare. You cannot prevent everything.
> Q.: **I take it, again, your basis of that belief would be that the fetus or embryo is an early human being?** (Emphasis added).
> A.: **Exactly. If it was a tooth, I would not worry about it.** (Emphasis added).

The court on hearing Dr. Lejeune and opposing expert witnesses, adopted Dr. Lejeune's views, and concluded as part of its finding that human life begins at conception, and rightly found the basis for its conclusion in the common law.* Other distinguished scientists concur with Dr. Lejeune's views. For example:

> Dr. Albert W. Liley, the Father of Modern Fetology: "We each began life as a single cell." [4]

In 1981 the U. S. Senate considered Senate Bill #158, the "Human Life Bill." Extensive hearings (eight days, 57 witnesses)

[4]www.vanderbilt.edu/SFL/lejeune_testimony.htm

*In consonance with the judiciary overriding the common law and scientific evidence in cases of abortion, the judge in the Tennessee State Supreme Court (Davis v. Davis, 842 S.W.2d 588, Tenn., 1992) on appeal overturned the decision of the lower trial and appeals courts. Many glaring errors in the reasoning of the court were disclosed, however. The Supreme

were conducted by Senator John East.[5] National and international authorities testified. After also hearing witnesses who disagreed [the pre-embryo doctrine would be in that category] with the experts holding the view that, "life begins at conception," the official Senate report, 97th Congress, S-158 recorded the following:

> Physicians, biologists, and other scientists agree that conception [they defined fertilization and conception to be the same] marks the beginning of the life of a human being—a being that is alive and is a member of the human species. There is overwhelming agreement on this point in countless medical, biological, and scientific writings [Report, Subcommittee on Separation of Powers

[5] www.abortionfacts.com/online_book...m_both/why_cant_we_love_them_both_11.asp.

Court used the pre-embryo doctrine to manufacture a reason for its decision. It discounted the testimony of Dr. Lejeune as a geneticist while adopting the opinions of obstetricians and gynecologists. However, a geneticist, working at the basic science level, is eminently qualified as an expert in defining when the life of a human being begins. The court brushed aside Dr. Lejeune's testimony by stating, "His testimony revealed a profound confusion between science and religion." In fact, it was the court which was profoundly confused. Dr. Lejeune's testimony was all science. He offered a moral conviction based on the scientific data he had collected. The court then, inexcusably and falsely, treated his expressed moral conviction as constituting a scientific fact presented by him. The court then proceeded to ignore Dr. Lejeune's scientific expert testimony, and, instead, adopted the opinions of the witnesses who promoted a pre-embryo view which holds that a pre-embryo is not to be considered a human being because it has not yet reached an embryo stage (A view discounted universally by credible national and international scientists as cited in the main body of the text of this Section). The court cited Robertson (*In the Beginning: The Legal Status of Early Embryos*, 76 Va. L. Rev. 437, 1990) for support for its decision. However, Robertson, in his article, presented no scientific foundation for holding that the pre-embryo is not an individual human being in existence. The facts contained therein only supports the view that the so-called pre-embryo is a new human being acting independently seeking out a synergistic relationship with its mother in order to complete its development. The pre-embryo position is simply arbitrarily chosen terminology, aimed at setting a time before which the embryo is said to exist, and serving to mislead an intellectually unsophis-

to Senate Judiciary Committee S-158, 97th Congress, 1st Session 1981, p. 7]. On pages 7-9, the report lists a "limited sample" of 13 medical textbooks, all of which state categorically that the life of an individual human begins at conception. Then, on pages 9-10, the report quotes several outstanding authorities who testified personally:

> Professor J. Lejeune, Paris: "Each individual has a very neat beginning, at conception."

---

ticated audience. As revealed in the main body of the text of this Section, it has been discredited by the scientific community and has found no favor among the learned geneticists and scientists in the field of genetics and embryology. "In rigorous ethical debate such arbitrary terminology, particularly if used to assign moral values, should be avoided."[5] It only finds acceptance in judicial and pro-abortion and anti-life circles as a pretended basis for eliminating the rights of the unborn.

Yet, all of the fuss about embryo versus pre-embryo is irrelevant in deciding the status of the beginning and rights of a human being because those issues have already been decided by controlling common law.

The fatal error in the reasoning of the Tennessee Supreme Court can be found in its statement (p. 596), "But, if **we have no** statutory or **common law precedents to guide us**, we do have the benefit of extensive comment and analysis of the legal journals." (Emphasis supplied). The court simply did not do its fundamental homework. Extensive binding common law precedents exist which decisively settle the points in question in the case, and the trial court correctly based its decision in finding that the right to life finds its basis in the common law.

A study of Section II of this treatise will demonstrate that personhood, life, and a human being all have their beginning at the time of conception. To declare a pre-embryo stage of existence in order to erase the recognition of life, personhood, and a human being as existing from the time of conception is an exercise in futility, because the common law has firmly addressed those contingencies and found all of those attributes exist from the time of conception. In Section II refer to the segments of Fleta, Pope Gregory IX, and Blackstone in which it is disclosed that firmly established immutable common law precedent holds that it is homicide to even artificially prevent procreation (contraception). **Thus, under existing controlling common law, the life of a human being is protected even dating from the time antecedent to the existence of the so-called pre-embryo stage of development.**

> Professor W. Bowes, University of Colorado: Beginning of human life?—"At conception."
>
> Professor H. Gordon, Mayo Clinic: "It is an established fact that human life begins at conception."
>
> Professor M. Matthews-Roth, Harvard University: "It is scientifically correct to say that individual human life begins at conception."
>
> **Those witnesses who testified that science cannot say whether unborn children are human beings were speaking in every instance to the value question rather than the scientific question. No witness raised any evidence to refute the biological fact that from the moment of human conception there exists a distinct individual being who is alive and is of the human species.** (Emphasis supplied).

Scientifically and objectively the case is closed, proving a person has his beginning with the formation of the zygote, and that abortion is a crime. It is simply raw judicial power that grants the privilege of killing without penalty. Future remedial efforts will have to be directed at disgorging this barbaric and tyrannical evil infecting our society and that of the world. Tolerance of the present day deviant thinking and savage conduct that permits abortion must stop. The present behavior of our country is like a family that inherits an exceptionally successful international business built by their very industrious and intelligent father, but because their sole interest consists in satisfying their insatiable appetites for every selfish wish and desire above everything else, they squander their largesse bringing the business to collapse and ruin.

There will be a repetition of many of these thoughts in the Sections to follow, as it is important, for purposes of clarity, to treat the subject separately in each Section as it applies to the particular topic or matter under discussion.

Section II

Ancient and Common Law Governing Abortion

INTRODUCTION

The foregoing discussion in Section I was devoted to the task of demonstrating what constitutes human life, a human being, and personhood. Thus, the human being within the uterus of the woman from the moment of fertilization until birth is a human life and person entitled to protection from being killed or injured. It is now worthwhile to identify the controlling principles under ancient and common law that confirm these truths, and the illicit nature of abortion, which constitutes the destruction of a human being. It is the common law that binds American jurisprudence in ruling on abortion issues. Ancient law is cited to demonstrate conclusively how longstanding and sweeping has been the recognition that abortion is a contemptible act, and to show how deviant and bizarre the recent rulings have been which, in only the past few years, have permitted this savage slaughter of human beings without penalty—even going so far as to call it a right!!

ANCIENT LAWS AND CUSTOMS

In the historical and apocryphal works, dating back thousands of years B. C., we find that human life was considered to be sacred from the moment of conception, and that prevention of that life or its destruction in utero was continuously condemned in an unbroken chain of laws and precedents comprising the common law inherited by America. Even in pagan, uncivilized, and barbaric societies this sacredness of human life was recognized.

In The Book of the Secrets of Enoch[1] which was translated from manuscripts of the Pseudepigraphal Group reveals the following teaching:

> And I swear to you, yea, yea, that there has been no man in his mother's womb, *but that* already before, even to each one there is a place prepared for the repose of that soul, and a measure fixed how much it is intended that a man be tried in this world.
>
> Yea, children, deceive not yourselves, for there has been previously prepared a place for every soul of man.

Comment:

This teaching would preclude even the use of birth-preventive measures as interfering with the preordained plans of God for each person brought into existence by him.

Pertinent Biblical citations which teach that the embryo is a human being from the earliest moment of conception, which thereby means that to kill that life in utero is murder:

> Before I formed you in the womb I knew you, and before you were born I consecrated you.[2]
>
> My frame was not hidden from you, when I was being made in secret, intricately wrought in the depths of the earth.[3]

The Didache, or The Teaching of the Twelve Apostles,[4,5] written in the time between 60 and 90 A. D., contains the following teaching:

> You shall not kill the embryo by abortion and shall not cause the newborn to perish.

The Epistle of Barnabas,[6,7] was probably written by St. Barnabas who was a companion and co-worker with St. Paul. Some have felt it was accomplished before the Epistle of St. Jude and those of St. John.[7] The work had been cited by many ancient Fathers including Clemens, Alexandrinus, Origen, Eusebius, and Jerome.[7] The pertinent passage in the Epistle pertaining to abortion is as follows:

> Do not kill a fetus by abortion, or commit infanticide.

Minucius Felix writing a work titled, *Octavius*,[8] perhaps in 166 A. D. penned the following:

> There are some women who, by drinking medical preparations, (by medicaments and drinks) extinguish the source of the future man in their very bowels, and thus commit a parricide before they bring forth.

Tertullian writing between 198 and 204 A. D.[9] spoke of the evil of abortion, to wit:

> In our case, murder being once for all forbidden, we may not destroy even the foetus in the womb, while as yet the human being derives blood from other parts of the body for its sustenance. To hinder a birth is merely a speedier man-killing; nor does it matter whether you take away a life that is born, or destroy one that is coming to the birth. That is a man which is going to be one; you have the fruit already in its seed.

Ancient and Early Common Law of England

Introductory Comments:

It is recognized that what is detailed in this section requires studious reflection. The recitation of facts are set forth in a purely documentary way to show with certainty that the common law outlawing abortion, under the common law of England, existed from the earliest times and continued uninterruptedly until 1967 when England changed (invalidly) her laws and the United States followed suit in 1973. The controlling principles are demonstrated by quoting actual documentation rather than simply citing references. The disclosure of the facts themselves establishes historical accuracy. It is important to recite, with scrupulous detail and precision, the holdings under the English common laws because America adopted those laws together with the precedents represented therein. Comments of the author are inserted to clarify, point out the significance of disclosures, or explain various relationships that exist.

The common law of Anglo-Saxon England had its roots from var-

ious groups which invaded the island in the earliest part of its history. Pre-Christian Anglo-Saxon society had virtually no documents and no books.[10] *The Catholic faith spread into Briton during the 2nd Century, and the British King, Lucius, was received into that faith by Pope Eleutherus,*[11] *who held the Papacy from 175-189 A. D. The Britons then accepted the Catholic faith, and held it peacefully until the time of Emperor Diocletian (284-305 A. D.),*[11,12] *who commanded the destruction of all Christian churches, which persecutions took place from 303-305 A. D.*[13]

However, during the reign of the Roman Emperor, Constantine, from 306-337 A. D. who was received into the Catholic faith, the religion enjoyed widespread acceptance in England. The Catholic faith was granted formal recognition as the religion of the Roman Empire [of which England at the time was a part] under Emperors Gratian (367-383 A. D.), Valentinian (364-375 A. D.) and Theodosius (378-395 A. D.).[14] *As a result, the spread of the Catholic faith to England expanded ever more widely. It was embraced by the early English kings who adopted and promulgated canons and laws of the Catholic Church as part of the laws of the kingdom, which thereby formed many of the earliest common laws inasmuch as all ordinances proceeding from the king and 'Witena-gemot,' whether of a secular or ecclesiastical character, are considered as Laws.*[15] *The Witan exercised judicial powers, becoming the highest court of law in the kingdom.*[16] *The king had no judicial powers separate from those of the court in which he sat. He was simply the presiding officer of that court, with executive powers to carry out its decrees.*[17] *During the earliest period, Catholic scholars, Catholic Church councils, writings of St. Augustine, and Catholic ecclesiastical rulings and decrees deeply influenced the laws which were developed and promulgated governing abortion. Those ordinances, canons, and rules of a strictly ecclesiastical nature constituted a proper, nay, an essential, part of the common law established by various kings enforcing compliance with the Catholic faith. The Catholic Church has always declared abortion to be evil and to be condemned, and it is not possible to adopt the Catholic faith without adherence to that tenet, and any modifications of its doctrines and decrees that the Catholic Church may issue from time to time in order to comply with more exacting knowledge as it becomes available. Only its dogmas remain unchangeable. The earliest common laws grew out of laws, canons, and penitential observances of the Roman Catholic Church, and the kings, in association with their 'witan,' continually fashioned their ordinances and laws to be in compliance with the moral teachings of that faith. Until the time of King Henry VIII, England was a Catholic country. Because the influence of the Catholic Church is fundamental in the development of the common law, out of necessity, repeated reference is made to this association throughout this treatise. Rather than simply citing this fact, repetitive disclosure of the various laws passed by the kings are quoted so as to set out accurately the historical background for this important element which impacted so vitally on the development of the common law of England, and ultimately adopted by America at its founding. A further purpose in repeatedly citing the various instances by detailed quotes of the actual laws and rulings of the early English kings is to show their steadfast, purposeful, and firm intent in promulgating their laws to be in accordance with the moral teachings of the Catholic Church, and to provide precise and accurate insight into those objectives untainted by any personal convictions or interpretations of the author.*

Another point that bears special attention is the view held in early Christian times as to the time ensoulment occurs. A point of view first advanced by Aristotle and later given favorable treatment by St. Augustine (354-430 A. D.) and St. Aquinas (1225-1274 A. D.) was that the person is not endowed with a soul until 40 - 60 (or even 90 days) after conception. For this reason, although abortion under the common law was always considered an evil, the penalty (often a penance) assigned was often of a lesser magnitude for an early abortion than if it took place at a later stage of pregnancy. This notion of an abortion being less serious if carried out in the early period of gestation, was also fostered by an interpretation arising on the basis of the Septuagint translation of Exodus 21:22-25 which passage refers to the situation of a fight between two men, as a result of which a pregnant woman loses her child. This early interpretation held that when the child is in the early stages of embryonic development, the penalty was a financial one, but if the child was in the later stages of development, the penalty was death.[18] *However, this passage from the Bible when properly translated from the original Hebrew carried a far more serious teaching than early theologians attached to it who had used the Septuagint version of the Old Testament. Inasmuch as the intent of the common law was to comply with this Mosaical teaching, it would, under the accurate translation, be required to find abortion at any stage of gestation, to be a crime for taking the life of the conceptus, and, that the father has a vested interest in the conceptus (See Hawkins abortion holding, infra, as well as Comment 3 thereto). Nonetheless, abortion at any stage of pregnancy was always considered a serious evil under the common law as will be seen in the disclosures that follow.*

As an introductory comment the author wants to call attention of the reader to the problem associated with the translation of the Latin word, anima, which can be translated as either spirit, life, or soul and the option selected in interpreting various texts must be understood in this context. ***However, the 40-day demarcation period established by the earliest precedents, has to apply to ensoulment, rather than quickening, because quickening occurs much later, usually about five months after conception.***

The relationship of the development of the common law of England to the teachings and doctrines of the Catholic faith can be appreciated by reviewing what various historical scholars on this subject had to say. As Plucknett observed:

> And finally, the Church brought with it moral ideas which were to revolutionise English law.[19]

The penances assigned by the Church for crimes committed were enforced by the secular government as part of their laws as these deterrences to commission of wrongdoing were viewed by the kings as a means of keeping the peace in the realm. Hence, the penitentials assigned by Archbishop Theodore or Archbishop Ecgbert or ordinances or rulings emanating from any other Catholic ecclesiastical source became a part of the Anglo-Saxon common law. The reader is referred to Oakley's detailed coverage of this subject, particularly Chapters V and VI. [20] *Selected excerpts from his text can be cited to advantage showing the special relationship that existed between the Church and State that gave rise to the common law finding its birth in the ecclesiastical penitentials, to wit:*

> At least until the growth of private jurisdictions, there were no distinct ecclesiastical courts separate from the secular ones; though there may have been "a system of church judicature with properly designated judges, and a recognised, though not well defined area of subject-matter in persons and things." The judicial matters of the Church were apparently transacted in the ordinary *gemots* of the hundred and of the shire.[21]
>
> To make the alliance between Church and State even closer, there were all-inclusive enactments by secular law, decreeing penance for sins in general and for violation of Church canons or of secular law. The terms of these laws make certain that they directly referred to ecclesiastical penance, rather than to money compensation to the Church, though this is sometimes mentioned in addition and is found in other sections. General, all-inclusive enactments requiring the performance of ecclesiastical penance for secular crimes are found in many prologues to pre-Norman laws, both in the Anglo-Saxon and Latin versions. The treaty between Edward the Elder and Guthrum states in the prologue: "And they established secular penal laws also, for that they knew that they could not otherwise guard against (much crime), *and many people would not otherwise subject themselves to religious penances, as they should do; and they established a secular compensation jointly for Christ and the king, wherever anyone should not wish to subject himself lawfully to churchly penance according to the bishop's command (or direction)".*[22] (Emphasis in original).
>
> In summary, the foregoing discussion has shown clearly that, with increasing thoroughness and severity, the pre-Norman kings constantly enforced penitential practices as supplementary aids to the suppression of disorder and crime. In doing so, they naturally did not replace with their own the authority over penance already vested in bishop and priest; but, undoubtedly, they strongly reinforced the actual performance of penance by the

threat of additional secular penalty, social ostracism, and personal disfavor at court. Pre-Norman England was very slowly converted, suffered several relapses of portions of the population into paganism, and may, at first, have offered some resistance to penance. As will be demonstrated, the general conditions in England, during the period in question, represented a fierce conflict between the powers of order and disorder, of crime and public peace. During that struggle both secular and ecclesiastical discipline cooperated in the process of socialising and civilising; and, in that cooperation, the effect of penitential measures was undoubtedly greater than has hitherto been supposed.[23]

Even Matthew Hale, legal scholar expressing anti-Papal sentiments, admitted the importance the Catholic Church had in the development of the common law by incorporating its moral teachings in those precedents:

Again, The growth of Christianity in this Kingdom, and the Reception of Learned Men from other Parts, especially from *Rome*, and the Credit that they obtained here, might reasonably introduce some *New* Laws, and antiquate or abrogate some *Old* ones that seem'd less consistent with the Christian Doctrines, and by this Means, not only some of the Judicial Laws of the *Jews*, but also some Points relating to, or bordering upon, or derived from the Canon or Civil Laws, as may be seen in those Laws of the ancient Kings, *Ina, Alphred, Canutus, etc.* collected by Mr. *Lambard*.[24]

To summarize:

The Church of England was the name given to that portion of the laity and clergy of the Catholic Church resident in England during the days of the Anglo-Saxon monarchy and during the history of England under William the Conqueror and his successors down to the time when Henry VIII assumed unto himself the position of spiritual and temporal head of the English Church. Prior to the time of Henry VIII, the Church of England was distinctly and avowedly a part of the Church universal. **Its prerogatives and its constitution were wrought into the fibre of the common law. Its ecclesiastical courts were recognized by the common law—the jus publicum of the kingdom—and clear**

recognition was accorded to the right of appeal to the sovereign pontiff; thus practically making the pontiff the supreme judge for England as he was for the remainder of Christendom in all ecclesiastical causes.[25] (Emphasis added).

* * * * *

King Ethelbert

The earliest written records of Anglo-Saxon common law come down to us from the time of King Ethelbert[26] who ruled as the king of Kent from 560 to 616 A. D. He was baptized by Augustine in 597 A. D. and established with the consent of his counselors a code of law inspired by the example of the Romans, the first of his laws being designed to protect those persons and doctrines he had embraced.[27]

* * * * *

Archbishop Theodore

The judgments/legal opinions applied to Penitents in England regarding abortion as recorded by Theodore, Archbishop of Canterbury, who served in this position from 668 to 690, were defined in his Penitentials as follows:

XVI. CONCERNING FORNICATION
OF [i. e., BY] LAY PEOPLE

§ 17. An unmarried woman committing immorality shall be a penitent for 3 years. If [she commits immorality] to the point of giving birth to a child, 3 whole years, and 2 years more lightly. If she kills it in the womb, 10 years; if after its birth, 15 years.[28]

XVIII. CONCERNING FORNICATION
OF [i. e., BY] CLERGY OR RELIGIOUS

§ 8. If, moreover, for the sake of concealing [the act], they kill children born in this way [i. e., illegitimately], there is indeed an ancient ruling which removes them from the church up to the time of their passing from this life, not only them, but even those who deal with them to expel the ones conceived in the womb. Nowadays there is a more humane rule, that they should do penance for 10 years, and should never go/be let off without any penance.[29]

Comment:

These laws identify the act of "killing" a child in the womb which ruling thereby bestows personhood on the one killed, and the penalty (she shall be a penitent for 10 years) is the same without distinction as to the period of gestation. Thus, abortion ("killing" in the womb) at any stage of pregnancy comports with what is designated as a crime in later jurisprudence. It also means that since only something alive can be killed, and when referencing a human pregnancy "the ***ones*** *conceived" can only be a person, both "life" and "person" were recognized by this law as pertaining to the in utero products of conception. Importantly, as stated in § 8 above, there was an even earlier—ancient, if you will—ruling punishing abortion more severely than the instant cited ruling.*

Continuing with the judgments/legal opinions of Archbishop Theodore:

XXI. CONCERNING MURDERERS/HOMICIDES

§ 3. Women who fornicate and kill their offspring, and those who deal with them to expel the ones conceived in the womb, an ancient ruling indeed removes from the church up to the time of their passing from this life. Now it is more humanely decreed that they should do a 10-year's penance.[30]

§ 4. A woman losing her offspring spontaneously before 40 days shall be a penitent for a year. But if [she loses the child] after 40 days[1], she shall [do penance] 3 years. But if after it shall have become quickened, [she shall do penance] as if she is homicide, that is, 10 years. But the case is a lot different if a woman in poverty does this because of the difficulty of nourishing the child, or if an immoral woman/fornicator [does this] for the sake of concealing her crime.[31]

§ 5. [2]If a woman shall have made an abortion purposely, of her own free will, she shall do penance 10 years.[32]

> [1] She shall do penance like a homicide.*
>
> > *homicida, the Latin word used in the original text covers any sort of killing of a person, murder or manslaughter.
>
> [2] Women who make an abortion before it has a soul shall do penance one year, or 3 forty-day fasts [120 days] or forty-days according to the nature of the fault; and afterward [after it has a soul], that is, after 40 days of accepting the seed she shall do penance as a homicide, that is, for 3 years on the fourth and sixth days of the week and, in addition, on the third day of the week during Lent. This is ruled according to the canons for a period of 10 years.

Comment:

Reference is again made in § 3 to an earlier ancient ruling providing an even harsher penalty for committing abortion. In most instances it appears that the 40-day period, marking a juncture between severe and less severe penalty, pertains to the time of ensoulment in accordance with the views espoused by the early Catholic Scholars such as St. Augustine. It is to be noted that during the early period of the common law this 40-day period marks a significantly earlier period in pregnancy triggering the crime as homicide than that later enforced. However, it seems that the penalty for abortion by an unmarried pregnant woman .consisted of that for homicide regardless of the stage of pregnancy (§ 4). Noteworthy, is the finding that in the early common law, even a spontaneous miscarriage after the 40-day period, carried a penalty "like" homicide (§ 4, fn1), and after quickening it carried a penalty "as if" it was homicide (§ 4). The change, beginning with Coke, reducing the crime of abortion from homicide to a great misprision after quickening, as will be seen later in this text, was only adopted as an accommodation to assist in obtaining indictments. In all of Archbishop Theodore's cited penitentials, punishment was meted out for abortion done at any stage of gestation with a harsher penalty if committed after 40 days from conception (i.e., the time of ensoulment), and the most severe (as per § 4) if carried out after quickening. The remedies applied only to modify punishment, to meet various circumstances, did not violate the principle that abortion at any stage of pregnancy is a crime. The established principles governing common law, as will be shown later by Coke, Blackstone, and the Dr. Bonham's Case, cannot be changed by any means whatsoever except to remedy defects (which would include modification of punishments) to enhance the particular common law in question. A precedent would have little meaning, in serving its legal purpose of stare decisis, if the principle it represents were not permanently fixed in place to guarantee that justice in the future would be meted out the same in all similarly situated cases. These laws of Archbishop Theodore, all listed under the category of homicide, as well as those later proclaimed by Archbishop Ecgbert, show clearly that early abortions did provide for indictment as a crime despite the mistaken beliefs of some of the early American courts.

A ruling against abortion that antedated the time of Archbishop Theodore was Canon XXI, adopted at the Council of Ancyra in 314 A. D., as follows:

> Concerning women who commit fornication, and destroy that which they have conceived, or who are employed in making drugs for abortion, a former decree excluded them until the hour of death, and to this some have assented. Nevertheless, being desirous to use somewhat greater lenity, we have ordained that they fulfil ten years [of penance], according to the prescribed degrees.[33]

The following points are of importance to note: 1. Canon XXI of the Council of Ancyra closely parallels Penitentials XVIII § 8 and XXI § 3 of Archbishop Theodore. 2. Archbishop Theodore, in both of those Penitentials, refers to "an ancient ruling" which carried a more severe penance for committing abortion than the ones he prescribed. 3. The Ancyra Canon speaks of "a former decree" carrying a more severe punishment for abortion than the one it adopted. 4. The more severe penance, mentioned by both Archbishop Theodore and the Council of Ancyra, was the same (excluded from the church until the time of death). 5. The penance for abortion adopted by the Council of Ancyra in 314 A. D. corresponds almost exactly with the harshest penance promulgated for the crime 800 years later by King Henry I (1100-1135)—which see, later in this Section. 6. Therefore, severe punishment for committing abortion, had to have existed before the time of the Council of Ancyra of 314 A. D. Thus, the common law emanating from Archbishop Theodore's Penitentials, and carried forward to the present time (although invalidly ignored by the U. S. Supreme Court in Roe v. Wade) actually found its inception at a date earlier than the Council of Ancyra in 314 A. D.

Another even earlier ruling condemning abortion was found in Canon 63 adopted by the Council of Elvira in 305 A. D., to wit:

> If a woman conceives in adultery and then has an abortion, she may not commune again, even as death approaches, because she has sinned twice.[34]

* * * * *

King St. Ine

King St. Ine (hereinafter, King Ine) ruled as the King of West Saxon from 688 to 725. He was a devout Catholic, and his law code reveals a Christian society so well established that some breaches of Church law met secular punishment; there are also passages which reveal the existence of high-status Britons under Saxon rule.[35] That the prelates of the Catholic Church were an important component of his council, and instrumental in the drafting of his laws is brought out in the preamble and first of the laws he enacted, to wit:

OF THE DOOMS OF INE

> I, Ine, by God's grace, king of the West-Saxons, with the counsel and with the teaching of Cenred my father, and of Hedde my bishop, and of Eorcenwold my bishop, with all my 'ealdormen,' and the most distinguished 'witan' of my people, and also with a large assembly of God's servants, have been considering of the health of our souls, and of the stability of our realm; so that just law and just kingly dooms might be settled and established throughout our folk; so that none of the 'ealdormen,' nor of our subjects, should hereafter pervert these our dooms.[36]

OF THE RULE OF GOD'S SERVANTS

> 1. First, we command that God's servants rightly hold their lawful rule. After that, we command that the law and dooms of the whole folk be thus held.[36]

Comment:

By this law King Ine declared the independent holdings of the Catholic Church have lawful and binding standing, being absorbed thereby into the common law of England. His law would permit accommodation to future new or changing of rules as might be promulgated by the Catholic Church.

* * * * *

King Wihtraed

The continued development of the common law in conformity with Catholic Church doctrine is depicted in the manner by which King Wihtraed, the king of Kent from 690-725, carried out the formulation of his laws, in his own words, to wit:

THE LAWS OF KING WIHTRAED.

THESE ARE THE DOOMS OF WIHTRAED, KING OF THE KENTISH-MEN

In the reign of the most clement king of the Kentish-men, Wihtraed, in the fifth year of his reign, the ninth indiction, the sixth day of Rugern, in the place which is called Bergham-styde, where was assembled a deliberative convention of the great men: there was Birhtwald archbishop of Britain, and the fore-named king; also the bishop of Rochester, the same was called Gybmund, was present; and every degree of the church of that province spoke in unison with the obedient people. There the great men decreed, with the suffrages of all, these dooms, and added them to the lawful customs of the Kentish-men, as it hereafter saith and declareth.[37]*

*Several of the ecclesiastical provisions in the enacted dooms appear to have been founded on the canons of the synod of Hertford, held in the year 673.[38]

* * * * *

Archbishop Ecgbert of York

Archbishop Ecgbert of York who served in this capacity from 735 to 766, laid down the following law codes:

30. A woman who expels the thing-conceived in her womb, and 40 days after the seed is received kills it, before it has been ensouled, shall fast 3 years as if [she were] a homicide, and in whatever [any?] week 2 days until evening, and the 3 designated fasts; if she loses/destroys she shall fast one year or the 3 designated fasts.[39]

A footnote to the above reads:

The same judgment shall be applied to women who make an abortion of their children: they shall be considered homicides [if an abortion is made] before the child is living,* or afterwards,

> that is 40 days after the seed has been received, and then they shall fast for three years, on the days of Wednesday and Friday and in the three designated fast-periods.[40]
>
> > *Undoubtedly, ensoulment, because of the relationship as a footnote to 30, above, and because of the follow-on part of the sentence.

In Book II of his Penitential Writings he prescribes:

> 2. If any woman shall have lost a child within her [womb], by means of a potion/drink or by any other means whatever, or then kills it after it is born, she shall fast 10 years—3 years on bread and water and 7 such as her confessor wishes mercifully to prescribe for her.[41]

In Book IV of his Penitential Writings he prescribes:

> 21. If any woman by her own potion/drink destroys her child within her by her own free will, or kills it by any other means whatever, she shall fast for 7 years, 3 on bread and water, and in the other [years] shall enjoy her [normal] food except for meat alone.[42]

Comment:

In No. 30 Archbishop Ecgbert explicitly speaks of "killing" the "thing" which means the Church has assigned "personhood" and life to "it," that is, before quickening has occurred. The Archbishop further identifies abortion at or after the 40-day period, even if the "thing" is not yet ensouled, as being homicide. In his footnote he very clearly makes the point that ***abortion is homicide whether or not the in utero child is ensouled (or "living"),*** *and treats the unlawful action as homicide or killing in all four of his declarations. This common law was and has been improperly ignored by later commentators, legal scholars, and courts. Archbishop Ecgbert is following the teachings of the Didache, St. Barnabas, and St. Clement in holding that abortion is unlawful at any stage of pregnancy. Archbishop*

Theodore, giving deferential consideration to the teachings of Augustine, assigned a less severe penance for an early as opposed to a late abortion. Archbishop Ecgbert assigned the same punishment for abortion carried out at any stage of gestation. This assignment of different punishments for the same crime was simply modifying a remedy of a common law precedent while retaining its principle, which is permissible. The principle enunciated in these cases is that abortion is a crime at any stage of gestation, and ***this established legal precedent has never changed nor can it be****,* *<u>and advanced medical and scientific knowledge only confirms the wisdom and accuracy of their rulings.</u> Although this established precedent of the common law principle governing abortion has remained unchanged, it has been ignored, misunderstood, or misstated by later legal authorities.*

Stated succinctly, abortion at any stage of gestation was established under the common law as constituting homicide, with either a lesser or the same punishment being meted out if the act was committed within the first 40 days after conception as opposed to the period beginning after 40 days from the time of conception. The other immutable precedents established were that "life" and "personhood" have their inception at the time of conception.

* * * * *

King Alfred

King Alfred reigned as King of Wessex from 871-901. The various tribes became united into a single nation during his reign and their varying customs had to become harmonized, so that the first laws of all England were written during his rule.[43] King Alfred was received into the Catholic faith, and during his rule, conformed his laws to harmonize with the teachings of the Catholic Church. The following excerpt from his laws is pertinent:

ALFRED'S DOOMS

> 18. If anyone, in strife [earnest effort or endeavor], hurt a breeding woman, let him make 'bot' [damages] for the hurt, as the judges shall prescribe to him. If she die, let him give soul for soul.[44]

The following law was also enacted by King Alfred in collaboration with his 'witan:'

OF SLAYING A CHILD-BEARING WOMAN.

> 9. If a man kill a woman with her child, while the child is in her, let him pay for the woman her full 'wer-gild,'* and pay for the child half a 'wer-gild,' according to the 'wer' of the father's kin.[45]
>
> > *The price at which every man was valued, according to his degree, which in the event of his being slain, was to be paid to his relatives, or his 'gild-brethren,' by the homicide or his friends, and which he was himself condemned to pay, if proved guilty of certain offenses specified in the laws.

Comment:

The payment of 'wer-gild' for the child in the womb documents the holding that an in utero human life was considered a human being or person without respect to the period of gestation. Also, it is important to recognize that payment of 'wer-gild' according to the 'wer' of the father's kin recognizes that rights flow to the father in cases of abortion for the vested interest he has in the unborn child.

King Alfred indicated the judicial scope accorded to the Church, for example, in dealing with the matter of oaths and 'weds' [pledges]:

> But if he pledge himself to that which it is lawful to fulfil, and in that belie himself, let him submissively deliver up his weapon and his goods to the keeping of his friends, and be in prison forty

days in a king's 'tun:' [Villa - a dwelling with the land enclosed about it] let him there suffer *whatever the bishop may prescribe to him*; and let his kinsmen feed him, if he himself have no food.[46] (Emphasis supplied).

* * * * *

King Edward

King Edward who reigned as king from 901 to 924 adopted the laws jointly agreed to previously by King Alfred and King Guthrum, as here expressed by King Edward:

AGAIN HIS, AND GUTHRUM'S, AND EDWARD'S.

These are the dooms which king Alfred and king Guthrum chose.

And this is the ordinance also which king Alfred and king Guthrum, and afterwards king Edward and king Guthrum, chose and ordained, when the English and Danes fully took to peace and to friendship; and the 'witan' also, who were afterwards, oft and unseldom that same renewed and increased with good.

This is the first which they ordained: that they would love one God, and zealously renounce every kind of heathendom. And they established worldly rules also for these reasons, that they knew that else they might not many controul, nor would many men else submit to divine 'bot' as they should: and the worldly 'bot' they established in common to Christ and to the king, wheresoever a man would not lawfully submit to divine 'bot,' by direction of the bishops.

2. And if anyone violate christianity, or reverence heathenism, by word or by work, let him pay as well 'wer,' [see definition, supra, for 'wer-gild,' with 'wer' having a similar meaning] as 'wite' [fine] or 'lah-slit,' [a fine for offenses committed by the Danes, for which the English were condemned in the 'wite'.] according as the deed may be.[47]

Comment:

It is clear that the laws and punishments are established "in common to Christ and to the king," documenting with certainty the validity of Catholic Church doctrine, canons, rules, and ordinances for inclusion among the laws comprising the common law of England. As revealed, the norms of conduct for society find their basis in laws in conformity with and defined by Christianity (i.e., the Catholic Church, since that was the only Christian religion of the time). Importantly, jurisdiction for punishment is ceded to the bishops. Thus, the canons, codes, or penitential obligations imposed by the Church became a part of the common law. This jurisdictional interrelationship existing between Church and king is shown by Law 4, thus:

OF INCESTUOUS PERSONS.

> 4. And concerning incestuous persons, the 'witan' have ordained, that the king shall have the upper, and the bishop the nether, unless 'bot' be made before God and before the world, according as the deed may be; so as the bishop may teach.[48]

King Edward made the following declaration while king:

> To all who are charged with the administration of public affairs I give the express command that they show themselves in all things to be just judges precisely as in the Liber Judicialis it is written; nor shall any of them fear to declare the common law freely and courageously.[49]

Comment:

Liber Judicialis was the Dome-book containing the usages and customs having the force of common law (the existence of which is explicitly recognized by King Edward's declaration) that was compiled by King Alfred (Alfred the Great). The Liber Judicialis confirmed the indispensable role the Catholic Church played in the development of the common law of England.

* * * * *

King Aethelstan

King Aethelstan assumed to the throne in 924. During his reign laws were established including those for London, following which the king wrote:

> This is the ordinance which the bishops and the reeves belonging to London have ordained, and with 'weds' confirmed, among our 'frith-gegildas,' as well 'eorlish' as 'ceorlish,' in addition to the dooms which were fixed at 'Greatanlea' and at Exeter and at 'Thunresfeld.' [50]

Comment:

This preamble shows how integrally the Catholic Church prelates were in formulating the laws of England.

Again, this same relationship is disclosed in a statement found in some manuscripts in connection with the passing of "Aethelstan's Laws," in which the king ends with this conclusory clause:

> All this was established in the great synod at 'Greatanlea:' in which was the archbishop Wulfhelm, with all the noble men and 'witan' whom King Aethelstan.......gather.[51]

* * * * *

King Edmund

King Edmund who was a brother of King Aethelstan began his reign in 940 and was assassinated in 946. During this time Odo was Archbishop of Canterbury from 934 to 958 and Wulfram was Archbishop of York from 939 to 952. Ecclesiastical and secular laws were established during Edmund's reign and from the records of a synod which the king convoked are found the following:

THE LAWS OF KING EDMUND.

ECCLESIASTICAL.

KING EDMUND'S INSTITUTES.

King Edmund assembled a great synod at London, during the holy Easter tide, as well of ecclesiastical as of secular degree. There was Oda archbishop, and Wulfstan archbishop, and many other bishops, meditating concerning the condition of their souls, and of those who were subject to them.[52]

OF HOMICIDE.

3. If any one shed a Christian man's blood, *let him not come into the king's presence*, ere he go to penance, as the bishop may teach him, and his confessor direct him.[53] (Emphasis supplied).

SECULAR.

OF BLOOD-SHEDDING.

4. Also I make known that I will not have to 'socn' in my household that man who sheds man's blood, before he has undertaken ecclesiastical 'bot,' and made 'bot' to the kindred, * * * * and submitted to every law, as the bishop shall teach him in whose shire it may be.[54]

Comment:

Here again, deference is shown by the king in sharing decision-making with the ecclesiastical authorities, and granting priority of jurisdiction to the Catholic clergy for evaluating the criminal's act and determining the appropriate punishment.

* * * * *

King Edgar

King Edgar reigned from 959-975. Inasmuch as the 'witan-gemots' in conjunction with the king established laws constituting common law, it is illuminating to read how these assemblages were responsible for enacting both the secular and ecclesiastical laws of the realm. For example, as disclosed by King Edgar in connection with the laws he enacted:

> OF 'GEMOTS.'
>
> 5. And let the hundred-'gemot' be attended as it was before fixed; and thrice in the year let a 'burh-gemot' be held; and twice, a shire-'gemot;' and let there be present the bishop of the shire and the 'ealdorman,' and there *both expound as well the law of God* ***as*** *the secular law.*[55] (Emphasis supplied).

Comment:

There is often a misunderstanding that ecclesiastical laws were separate from secular laws in forming the body of common law. This excerpt shows how the 'witan' included the Church prelates (Only Catholic at this time), and that ecclesiastical laws adopted by the king were as binding under common law as the secular laws.

This association is again depicted by the following enactments:

> HERE IS THE ORDINANCE OF KING EDGAR.
>
> This is the ordinance that king Edgar, with the counsel of his 'witan,' ordained, in praise of God, and in honour to himself, and for the behoof of all his people.
>
> 1. These then are first: that God's churches be entitled to every right; and that every tithe be rendered to the old minster to which the district belongs; and that be then so paid, both from a thane's 'in-land,' and from 'geneat-land,' so as the plough traverses it.[56]

CANONS ENACTED UNDER KING EDGAR.

> 10. If a woman destroys her child within herself, or after it is born, by potions or whatever other means, she shall fast 10 years, 3 on bread and water, and 7 such as her confessor mercifully prescribes for her, and until she repents/does penance.[57]

Comment:

The Canons enacted under King Edgar carry the force of common law, because, as he stated, the 'gemot' expounded, "as well the law of God as the secular law." It is interesting to note how closely the law against abortion enacted by King Edgar parallels the decree of Archbishop Ecgbert in Book II of his Penitential writings after an elapsed time of approximately 200 years. Here, abortion is unlawful at any stage of gestation. A child in the Oxford English Dictionary is defined as "The unborn or newly born human being; foetus, infant. App. Originally always used in relation to the mother as the 'fruit of the womb,'" further confirming that this law of Edgar conforms to Archbishop Ecgbert's decree.

* * * * *

King Ethelred

Ethelred, son of King Edgar, succeeded to the throne of his father in the year 978 and died in 1016. Among the Laws of King Ethelred are the following:

IN THE NAME OF THE LORD,

IN THE YEAR OF THE LORD'S INCARNATION 1008.

> This is the ordinance that the king of the English, and both the ecclesiastical and lay 'witan,' have chosen and advised: [58]
>
> 1. **This then is first: that we all love and worship one God,**

and zealously hold one Christianity, and every heathenship totally cast out: and this we all have, both with word and with 'wed,' confirmed; that, under one kingship, we will observe one Christianity. And the ordinance of our lord and of his 'witan' is; that just law be set up, and every unlawfulness carefully abolished; and that every man be regarded as entitled to right; and that peace and friendship be lawfully observed, within this land, before God and before the world.[58] (Emphasis added).

3. And the ordinance of our lord and of his 'witan' is; that Christian men, for all too little, be not condemned to death: but in general let mild punishments be decreed, for the people's need; and let not for a little God's handywork and his own purchase be destroyed, which he dearly bought.[58]

11. And let God's dues be willingly paid every year: that is, plough-alms, xv. days after Easter, and a tithe of young by Pentecost, and of earth-fruits by Allhallows' mass, and Rome-'feoh' by St. Peter's mass, and light-scot thrice in the year.[59]

14. And let all St. Mary's feast-tides be strictly honoured; first with fasting, and afterwards with feasting: and at the celebration of every apostle, let there be fasting and feasting; except that on the festival of St. Philip and St. James we enjoin no fast, on account of the Easter festival.[60]

16. And the 'witan' have chosen, that St. Edward's mass-day shall be celebrated over all England on the xv. kal. April.[60]

31. And if any one anywhere commit 'forsteal,' or open opposition to the law of Christ or of the king; let him pay either 'wer,' or 'wite,' or 'lah-slit,' always according as the deed may be: and if he resist against right, by any violation of the law, and so act that he be slain, let him lie uncompensated to all his friends.[61]

34. **It is the duty of us all to love and worship one God, and strictly hold one Christianity, and totally cast out every kind of heathenism**.[61] (Emphasis added).

Comment:

King Ethelred's 1st and 34th laws show a continued commitment to the holdings of the Catholic faith, which is the one Christianity of which he speaks. Laws 11, 14, and 16 confirm the deference his laws pay to observances of the Catholic Church.

Law 3, which prescribes mild punishment, may account for the future development of more lenient punishments for the crime of abortion.

The Council of Enham held during King Ethelred's rule adopted the following ordinance:

COUNCIL OF ENHAM.

OF THE ORDINANCES OF THE 'WITAN.'

> 1. These are the ordinances which the councillors of the English selected and decreed, and strictly enjoined that they should be observed. And this then is first,—The primary ordinance of the bishops—that we all diligently turn from sins, as far as we can do so, and diligently confess our misdeeds, and strictly make 'bot,' and rightly love and worship one God, and unanimously hold one Christianity, and diligently eschew every heathenism, and diligently promote prayer among us, and diligently love peace and concord, and faithfully obey one royal lord, and diligently support him, with right fidelity.[62]

Comment:

This Law of King Ethelred unambiguously declares that England must hold one Christianity (The Roman Catholic faith), which also means its laws regarding abortion, must conform to the teachings of the Catholic Church on the subject. In and of themselves, these repeated declarations of the allegiance of England to the Catholic faith became a part of the common law.

Among laws adopted by King Ethelred in 1014 is the following:

> 24. And wise were also in former days those secular 'witan' who first *added secular laws to the just divine laws*, for bishops and consecrated bodies; and reverenced, for love of God, sanctity, and the sacred orders; and God's houses and God's servants firmly protected.[63] (Emphasis supplied).

Comment:

This law shows the secular laws were not in conflict with the divine laws, and the king recognized the divine laws as just. Secular laws were <u>*added*</u> *to the just divine laws but did not replace or usurp them.*

Other laws passed in 1014 were:

> 26. If a mass-priest become a homicide, or otherwise flagrantly commit crime, let him then forfeit both his order and his country, and be an exile as far as the pope may prescribe to him, and strictly do penance.[64]

> 38. And then was separated what was before in common to Christ and the king in secular government; and it has ever been the worse before God and before the world: let it now come to an amendment, if God will it.[65]

Comment:

Law 26 adopted by the king after counsel with his 'witan' shows the ceding of jurisdiction to the Catholic Church in a criminal matter under established common law.

Law 38 prescribes an amendment to correct a practice which had developed (in contravention of established common law) dividing the jurisdiction between secular and ecclesiastical matters. This law was also adopted by the king in counsel with his 'witan,' thereby correcting an erroneous application of the common law and to ensure that the actual common law would be recognized.

* * * * *

King Cnut

Cnut, King of Denmark, succeeded King Ethelred on the throne and became monarch of all England on the death of Edmund Ironside in 1017 and died in 1035.

Among the laws adopted during his reign are the following:

THE LAWS OF KING CNUT.

ECCLESIASTICAL.

> This is the ordinance that king Cnut, king of all England, and king of the Danes and Norwegians, decreed, with counsel of his 'witan,' to the praise of God, and to the honour and behoof of himself: and that was at the holy tide of Mid-winter, at Winchester.[66]
>
> 1. That then is first, that, above all other things, they should ever love and worship one God, and unanimously observe one Christianity, and love king Cnut with strict fidelity.[66]
>
> 2. And to hold in 'grith' and in 'frith,' and frequently to seek God's churches, for the salvation of souls and the behoof of ourselves. Every church is by right in Christ's own 'grith,' and every Christian man has great need that he show great reverence for that 'grith;' because God's 'grith' is of all 'griths' the most excellent to merit, and the best to preserve, and next thereto, the king's....etc.[66]
>
> 21. And we very earnestly instruct all Christian men, that they ever love God with inward heart, and diligently hold orthodox Christianity, and diligently obey the divine teachers, and meditate on and inquire into God's doctrines and laws, oft and frequently, for their own behoof.[67]

23. And we instruct, that every one shield himself very carefully against deep sins and diabolical deeds at every time; and that he very carefully make 'bot,' by counsel of his confessor, who, through impulse of the devil, has fallen into sins.[68]

Comment:

Law 1 directs that they (English, Danes, and Norwegians) unanimously observe one Christianity (the Catholic faith), and Law 21 reaffirms that they are instructed to diligently obey the divine teachers. Thus, the laws of the Catholic Church become the common law of the realm.

SECULAR

---

2. And we instruct, that though any one sin and deeply foredo himself, let the correction be regulated so that it be becoming before God and tolerable before the world. And let him who has power of judgment very earnestly bear in mind what he himself desires, when he thus says: 'Et dimitte nobis debita nostra, sicut et nos dimittimus.' And we command that Christian men be not, on any account, for altogether too little, condemned to death: but rather let gentle punishments be decreed, for the benefit of the people; and let not be destroyed for little God's handy-work, and his own purchase which he dearly bought.[69]

11. And be it constantly inquired, in every wise, how counsel may most especially be devised for the benefit of the nation; and orthodox Christianity most exalted, and unjust laws most diligently abolished: because through that it shall turn to some good in the country, that injustice be put down, and justice loved, before God and before the world. Amen.[70]

Comment:

The laws of King Cnut affirm the supremacy of the laws of God (to which the ecclesiastical laws pertain), that his subjects hold to the orthodox Christianity (the Catholic Church) and that his subjects obey the divine teachers. This means they must observe the teachings of the Catholic Church (and as prescribed and modified at any given time in the future), on abortion as a matter of ecclesiastical law which has been made a part of the common law of England, and that his subjects are subject to the penalties ('bot') levied by the ecclesiastical authorities (e.g., the confessors). King Cnut also adopted the position of king Ethelred that punishments be gentle, which perhaps provided an impetus for also reducing the penalty inflicted for abortion.

The laws of King Cnut closely mirror those of King Ethelred, particularly regarding ecclesiastical matters.

* * * * *

King St. Edward the Confessor

King St. Edward the Confessor's reign extended from 1042 to 1066.[71] He prepared digests of the laws of the realm; King Alfred's Dome-Book or "Liber Judicialis" survived in the improved code of Edward the Confessor.[72]

* * * * *

King William the Conqueror

King William the Conqueror ruled over England from 1066 to 1087. No definite break with the English law took place during this time.[73] However, under his rule he recognized the Ecclesiastical Courts of the Catholic Church as being entitled to jurisdiction in England in matters spiritual, and these church courts interpreted this recognition they were accorded very broadly in dealing with the administration of their canon laws, and concerning the matters in which they held jurisdiction, our law is still deeply colored by the doctrines of the canon law.[74] The following declaration was set forth by King William the Conqueror:

IV.
CARTA WILLELMI.

[1]W[illiam] by the grace of God King of the Angles [English], to R. Bainard & G. de Magnavilla and P. de Valoines, and the rest of my faithful people from Essex and from Hertfordshire and from Middelsex, greetings. Be it known to you all and to the rest of my faithful people who reside in England, that the episcopal [i.e., ecclesiastical] laws, which are not well[-done, i.e., proper, suitable], nor according to the teachings/laws of the holy canons, right up to my own times in the kingdom of the Angles/the English, by the common council and judgment of the archbishops [2]and bishops and the abbots, and of all the princes/rulers of my kingdom, I have judged must be corrected. For this reason, I command and give the order by my royal authority, that no bishop or archdeacon may hold ?or reach a judgment any longer in the ?civil courts? nor shall they bring any case which pertains to the regimen/system-of-rules of souls to the judgment of secular men; but whoever shall be called up according to the ecclesiastical laws for whatever reason/case or fault shall come to the place for this [court-case] which the bishop shall choose or name, and there shall answer concerning his case or fault, and not according to [civil law?], but according to the canons and church laws, shall make restitution to God and his [the defendant's] bishop. But if anyone, puffed up/elated with pride, scorns or refuses to come to the episcopal [ecclesiastical] judgment [=court case] let him be summoned once, a second, & a third time; but if even then he doesn't come for correction, he shall be excommunicated; and if he wrecks/subverts the process for making restitution for this, the strength and justice of the king or his stand-in/representative shall be brought to bear. However, that person who refuses to come when called into the church court for each and every time he is called up, he shall make amends. Moreover, I refuse and forbid by my authority that any representative or presider except a representative (minister) of the crown, or any layman, shall interfere/interpose himself concerning the laws which pertain to the bishop, nor shall any layman bring any other person to trial without the judgment/judicial

procedure of the bishop. Indeed the judgment/trial shall be carried on in no place except that of the bishop himself, or in that place which the bishop determines for it.[75]

> [1]William by the grace of God King of England to his associates, representatives, and all, Frenchborn and Englishborn, who hold lands in the bishopric of Bishop Remegius, greetings.[75]
>
> [2]..of my own and of the rest of the bishops.[75]

Comment:

This order promulgated by King William demonstrates the continued prime position enjoyed by the ecclesiastical courts in establishing the laws of England. Without break, the prior existing common law was continued during his reign and beyond. The primacy of the Catholic Church courts ensured that the abortion laws continued intact as they had from the foundation of England dating back to the Anglo-Saxon period.

* * * * *

King Henry I

King Henry I ruled from 1100 to 1135. In 1100 he granted a Charter of Liberties. Notable among the provisions is the following:

> 13. I restore to you the law of King Edward with those amendments introduced into it by my father with the advice of his barons.[76]

Comment:

It has been felt by some that this Charter may have served as the model for the Magna Charta of 1215.

King Henry I also enacted the following laws:

§ 14. If a pregnant woman is killed, and the child in her lives, for each one the full penalty-money is to be paid. If [the child] is not yet alive, half the penalty shall be dismissed for the parents on the father's side. But concerning the manbota of either one or the other, it shall go to the lord by right.[77]

§ 16. Women who commit fornication and kill their offspring, and those who deal with them to dislodge the conceived one from the womb, by an ancient decree, are removed from the Church to the end of their lives, now are judged more mercifully, shall do penance for 10 years. If a woman of her own free will loses an [unborn] child before 40 days, she shall do 4 years penance; if [she loses the child] after it is ensouled she shall do 7 years penance.[78]

Comment:

Law No. 16 concerning abortion, passed in the 12th Century, is essentially a duplicate and a continuation of that adopted approximately 500 years previously by Archbishop Theodore (XXI, § 3) and approximately 800 years previously by the Council of Ancyra. It is interesting to note that there has always been and continues to be a severe penalty for abortions carried out before the time the conceptus is considered ensouled. This means abortion has continued to be recognized from the very earliest times as a severe crime from the moment of conception onwards. The reference to the pre-40 day product of conception as "child" rather than "thing" illustrates the personhood attached to the being at the time of conception. Although the penalty prescribed for abortions committed before the 40 day period is less than that committed at or after the 40 day period (change of punishment is allowed under common law practice so long as the principle embodied in the precedent is not destroyed) ***there is no disclosure of a change in the view expressed by Archbishop Theodore 500 years previously and by the Council of Ancyra 800 years previously that abortion at either stage invoked the punishment meted out for homicide.***

Law No. 14 applies to a situation in which the intent of the criminal act is to kill the woman, who happens to be pregnant, as opposed to the intent being to kill the fetus, which would constitute abortion. The "not yet alive" undoubtedly refers to "detection of life" (i.e., quickening) as providing a method to determine that the fetus was not already dead when the killing of the woman occurred (Later discussion in Section III will disclose that the authorities admitted it presented them a real problem in trying to determine whether in utero life was snuffed out by the killing or the woman was already carrying a dead fetus according to the understanding of the physiology of gestation existing in medieval times.). This interpretation is necessary and correct which harmonizes the common law holdings that personhood and life begin at conception, and as those principles are applied to abortion. The Doom of King Alfred (No. 9) dealing with the same kind of situation would be controlling by the precedent it set, and since the principle of a common law precedent cannot be abrogated, the present interpretation given to King Henry I's law is again necessary and would be considered to be a remedy to the prior law without destroying its principle. Once again, the common law father's rights attaching even to the embryo (at all stages of gestation) are reaffirmed, and inasmuch as this principle of the common law has been established it can never be changed except to enhance it by remedying defects; it can never be abrogated.

King Henry I also enacted law, **LXXVII**, during his reign concerning the status of a freeborn (child) or slave.[79] In discussing this law, The Commissioners on the Public Records of the Kingdom [England] made the following observations:

> If a woman was condemned to slavery while in a state of pregnancy, the child was born free. Th. P. xvi. 33. *n.* 1. On the other hand, if a female slave was manumitted after she became pregnant, her child was born a slave. Ecg. Conf. c. 25.[80]

> It appears also that the principle laid down by Bracton and Fleta, of deciding on the *status* of the child from the *status* of the mother at the time of its conception, is as old as the Penitentials of Theodore and Ecgbert.[80] (Emphasis in original).

Comment:

The abbreviations of Th. P. and Ecg. Conf. refer to the Penitentials of Archbishop Theodore and Confessionale of Archbishop Ecgbert, respectively, and those principles they laid down dated back to the 7th and 8th Centuries, respectively. This documented historical fact establishes that "personhood" was recognized as beginning from the time of conception dating back to the earliest times in the development of the common law. It will be seen later in this treatise that, indeed, both Bracton and Fleta reaffirm this principle of the common law. Only a "person" can inherit a status as a free child or slave.

* * * * *

King Stephen

The reign of King Stephen (1135-1154) was often characterized as "the Anarchy," so great were the disorders which filled it attendant upon the disputed title to the Crown.[81] Nothing changed in the common law pertaining to abortion during his reign.

* * * * *

King Henry II

King Henry II ruled England from 1154 to 1189. Because of the anarchy that existed during the reign of King Stephen, King Henry II promulgated the Constitutions of Clarendon in 1164 which was a "remembrance or recognition of a certain part of the customs, liberties, and dignities of his [King Henry II] predecessors, that is to say of King Henry his grandfather and others, which ought to be observed and held in the kingdom." [82]

Comment:

By this Constitution of Clarendon, those laws pertaining to abor-

tion and personhood enacted in King Henry I's reign (see, supra) were repromulgated by King Henry II (which of necessity must remain unchanged anyway because the principle of established common law precedent cannot be abrogated, as will be seen in later discussions in this treatise).

* * * * *

The Magna Charta

The Magna Charta was realized in 1215.[83] It came into existence because, while King John (1199-1216) claimed royal prerogatives of his ancestors, his spiritual and temporal lords sought guarantees, once again, to prevent the anarchy of King Stephen's reign from recurring. The following extract from the preamble to that document is relevant:

> John, by the grace of God, king of England, lord of Ireland, duke of Normandy and Aquitaine, and count of Anjou, to the archbishops, bishops, abbots, earls, barons, justiciars, foresters, sheriffs, stewards, servants, and to all his bailiffs and liege subjects, greeting. Know that, having regard to God and for the salvation of our soul, and those of all our ancestors and heirs, and unto the honor of God and the advancement of holy church, and for the reform of our realm, by advice of our venerable fathers, Stephen archbishop of Canterbury, primate of all England and cardinal of the holy Roman Church, Henry archbishop of Dublin, William of London, Peter of Winchester, Jocelyn of Bath and Glastonbury, Hugh of Lincoln, Walter of Worcester, William of Coventry, Benedict of Rochester, bishops; of master Pandulf, subdeacon and member of the household of our lord the Pope, of brother Aymeric (master of the Knights of the Temple in England), and of the illustrious men William Marshall earl of Pembroke, William earl of Salisbury, William earl of Warenne, William earl of Arundel, Alan of Galloway (constable of Scotland), Waren Fitz Gerald, Peter Fits Herbert, Hubert de Burgh (seneschal of Poitou), Hugh de Neville, Matthew Fitz Herbert, Thomas Basset, Alan Basset, Philip d'Aubigny, Robert of Roppesley, John Marshall, John Fitz Hugh, and others, our liegemen.

Comment:

Here it is seen that this document is carried out by the advice of the Catholic prelates with one of the purposes being for "the advancement of [the objectives of] holy church," and thereby to reform the realm.

The two articles most pertinent to the matter at hand are the first (1) and the last (63), which are as follows:

> 1. In the first place we have granted to God, and by this our present charter confirmed for us and our heirs for ever that the English church shall be free, and shall have her rights entire, and her liberties inviolate; and we will that it be thus observed; which is apparent from this that the freedom of elections, which is reckoned most important and very essential to the English church, we, of our pure and unconstrained will, did grant, and did by our charter confirm and did obtain the ratification of the same from our lord, Pope Innocent III, before the quarrel arose between us and our barons: and this we will observe, and our will is that it be observed in good faith by our heirs for ever. We have also granted to all freemen of our kingdom, for us and our heirs for ever, all the underwritten liberties, to be had and held by them and their heirs, of us and our heirs for ever.
>
> 63. Wherefore it is our will, and we firmly enjoin, that the English Church be free, and that the men in our kingdom have and hold all the aforesaid liberties, rights, and concessions, well and peaceably, freely and quietly, fully and wholly, for themselves and their heirs, of us and our heirs, in all respects and in all places for ever, as is aforesaid. An oath, moreover, has been taken, as well on our part as on the part of the barons, that all these conditions aforesaid shall be kept in good faith and without evil intent. Given under our hand—the above-named and many others being witnesses—in the meadow which is called Runnymede, between Windsor and Staines, on the fifteenth day of June, in the seventeenth year of our reign.

Comment:

This historic document ensures the Catholic Church will be forever guaranteed to be protected in its moral holdings in England and those nations, like America, which accept the common law of England. These rights were violated when the U. S. Supreme Court ruling prohibited Catholic Hospitals from refusing to permit abortions. King Henry VIII also violated those guarantees, by the laws he laid down stripping the Catholic Church of its authority and jurisdiction, contrary to the perpetual guarantees of the Magna Charta. However, as we shall see from Bracton, Fleta, Coke, Hawkins, and Blackstone, England continued to hold the common law rulings outlawing abortion at every stage of gestation. The lesson to be learned here, however, is that when a nation breaks from adherence to the common law to which it subscribes, the inevitable result is anarchy exemplified by the way England lived under the rule of Henry VIII, levying severe penalties on those Catholics who professed their faith—all in violation of the guarantees promised in the Magna Charta. This anarchy then led to tyranny resulting, for example, in priests being condemned to death or life imprisonment for celebrating Mass. This path is now being traveled in America as we see the government applying punitive measures to the unborn and those individuals who, from a proper moral and legal standpoint, seek to protect those lives from abortion under the guarantees afforded by common law precedent. This tyranny is predictably spilling over into other walks of life where oppressive measures are being taken by the federal government impinging on and restricting freedoms and liberties of its citizens. Inasmuch as all rights have their foundation in the right to life [E.g., cf. Furman v. Georgia, 408 U. S. 238,290 (1971)] this development follows as a necessary consequence.

* * * * *

King Henry III

King Henry III who reigned from 1216 to 1272 professionalized the administration of justice.[84] The royal party had thought the king's relation to the provisions of the Magna Charta as that of a signatory only to a treaty, bound only by his express consent, but pressures from the nation's people persuaded him to re-issue the Great Charter which hardened into law the concept of the supremacy of the English constitution (the Magna Charta). In 1225 he re-issued the Magna Charta (and confirmed by King Edward I in 1297), incorporating thereto the following assurances:

> (1) FIRST, THAT WE HAVE GRANTED TO GOD, and by this present charter have confirmed for us and our heirs in perpetuity, that the English Church [the Catholic Church] shall be free, and shall have its rights undiminished, and its liberties unimpaired. That we wish this so to be observed, appears from the fact that of our own free will, before the outbreak of the present dispute between us and our barons, we granted and confirmed by charter the freedom of the Church's elections—a right reckoned to be of the greatest necessity and importance to it—and caused this to be confirmed by Pope Innocent III.[85]

Comment:

This charter was confirmed in 1297 by King Edward I. England observed the laws of the Catholic Church until the 16th century. By then the common law governing abortion was firmly set in place. America, which borrowed the common law from England, became bound by those decisions as prescribed in the issue and re-issues of the Magna Charta.

* * * * *

Pope Gregory IX

In 1234 Pope Gregory IX promulgated his Bull, Rex Pacificus, containing Si Aliquis, perhaps an early conciliar text.[86] Si Aliquis contains the following ruling regarding contraception and abortion:

> If anyone for the sake of fulfilling lust or in meditated hatred does something to a man or a woman, or gives them to drink, so that he cannot generate, or she conceive, or offspring be born, let it be held as homicide.[87]

Comment:

This ruling condemns both contraception and abortion, and with respect to the crimes—which are identified as homicide for both contraception and abortion—becomes part of the controlling common law of England. The wrongdoing in the case of abortion applies to any stage of pregnancy. The five book compilation composed by St. Raymond of Pennafort, in which this work appeared, was ordered by Pope Gregory IX to form a new canonical collection destined to replace all former collections.[88] *Pope Gregory IX's promulgation of St. Raymond's work gave the compilation binding force of law.*[89] *The Decretals of Pope Gregory IX, which includes Si Aliquis,*[90] *was cited by Blackstone as a part of the common law of England.*[91]

* * * * *

Henry de Bracton

Henry de Bracton who died in 1268 was a judge on the King's Bench from 1247-50 and 1253-57. His textbook, *De Legibus et Consuetudinibus Angliae [On the Laws and Customs of England]*, was written some time between 1220 and 1230, brought up to date about 1250, and was first printed in 1569. Regarding abortion we find the following in his treatise:

> If one strikes a pregnant woman or gives her poison in order to procure an abortion, if the foetus is already formed or quickened, especially if it is quickened, he commits homicide.[92]

Comment:

In the earliest days of the formation of the common law, Arch-

bishop Theodore spoke of "killing" the child (Penitential XVI, § 17), without regard to stage of pregnancy, and "killing" of a child equates with homicide. Archbishop Ecgbert assigned the degree of crime of abortion as homicide at any stage of gestation. The Council of Ancyra in 314 A. D. levied the penalty for abortion to be that assigned for homicide by the early church. Now, approximately a thousand years later, Bracton discloses that abortion is still considered homicide at any stage of pregnancy. All other intervening rulings are compatible with this view, although punishment was sometimes of less severity if it took place prior to ensoulment. Bracton does not distinguish a difference in degree of severity of the crime depending on the stage of gestation. The common law on this point, therefore, has become well settled by this time. The phrase, "especially if it is quickened" is compatible with common law assigning a greater degree of gravity to a late stage as opposed to an early stage abortion, but, nonetheless, both are homicide. This flexibility in establishing punishment for any crime is in accordance with established common law allowing for mitigating and other circumstances that exist surrounding the commission of the crime—e.g., how much deliberation was given to carrying out the act?, how certain is provability of the act?, etc. The law as espoused by Bracton is very compatible with the holding of Si Aliquis.

Another condemnation of abortion as being a crime, considered alongside castration, the sentence for which may be capital punishment or permanent exile with the forfeiture of all property, is set forth in the following reference in Bracton:

> If anyone forcibly interferes with a woman's internal organs in order to produce abortion, he is liable.[93]

Comment:

The foregoing is set out in the section, "Of Pleas of the Crown," dealing with crimes, and is considered in connection with castration, reciting that in those instances in which a man is castrated it is con-

sidered a crime even if done with the consent of the person castrated. It would appear, with its disclosure as made, that this same imperative and penalty would be assigned to abortion. Furthermore, since no caveats are appended to the criteria to delimit the culpability in the performance of abortion, the law must be interpreted to apply universally to all cases in which an abortion is carried out on a woman whether with or without her consent. And this law cited by Bracton, and apparently overlooked by abortion commentators leaves no doubt that abortion carried out at ***any stage of pregnancy*** *is a crime.*

Bracton also defines "person" as understood under the common law. This point is dealt with in his chapter, "Of Persons." The rights of the person, as he observes, extend from the time of conception, thus:

> So too if he is born of a free mother and an unfree father out of wedlock. It suffices [for him to be free] that the mother, though she [afterwards] was made a bondswoman [by the marriage], is free either at the time the offspring is conceived or at the time it is born, or at least at some time during the interval, for the misfortune of the mother ought not to injure him who is in her womb.[94]

Comment:

The averment is often put forth that the in utero being has no right to life because he is not a "person." The untenability of that position has been addressed with certitude by Bracton, Fleta (which see, infra), and, as previously mentioned under the Laws of King Henry I, it has been recognized to be the common law since at least the time of Archbishop Theodore in the 7th Century and Archbishop Ecgbert in the 8th Century that "personhood" exists from the moment of conception. It will be well to keep this information in mind concerning longstanding and firmly fixed precedents recognizing that life and personhood exist from the time of conception, and will be found to be pertinent in the Section III discussion of personhood in analyzing the Roe v. Wade decision. One cannot have personhood without having life. That is, personhood cannot exist without coexisting life being present. Human life and personhood coexist.

* * * * *

Fleta

In Fleta, an ancient law book written soon after 1290[95] by a learned English judge, the following representation of the common law on abortion is found:

> He, too, in strictness is a homicide who has pressed upon a pregnant woman or has given her poison or has struck her in order to procure an abortion or to prevent conception, if the foetus was already formed and quickened, and similarly he who has given or accepted poison with the intention of preventing procreation or conception. A woman also commits homicide if, by a potion or the like, she destroys a quickened child in her womb.[96]

Comment:

Here is recorded that not only is abortion a homicide, <u>but that it is also a homicide to use poison or force to prevent conception</u>. Thus, again, contraception under the common law is a homicide. This reference is also later cited by Coke who in turn is cited by Blackstone as a controlling precedent. The crime of abortion constitutes homicide if the in utero being is formed and quickened. Inasmuch as a fetus cannot be quickened without being formed, the use of the word, "formed" would be redundant unless the "and" were meant to be interpreted as "or." Also, being silent on what degree the crime would be assigned for abortions carried out in the pre-quickened stage of pregnancy leaves the antecedent precedents established for this period of gestation as the governing law. Artificial birth control is classified as homicide, so pre-quickening abortion can be no less severe a crime.

Because the early Anglo-Saxon laws found their model in the moral codes of the Christian religion, this condemnation of contraception is clearly in accord with the common law of England drawn from the lesson taught by the example of Onan (Gen:38, 8-10) who received capital punishment for the sin of coitus interruptus. The Old Testament was adopted by the Roman Catholic Church, so the lesson taught by this Biblical text is an integral part of its faith and thereby is incorporated in the common law of England. Therefore, the law prohibiting contraception disclosed by Fleta and Pope Gregory IX's, Si Aliquis, finds a sure basis for its existence in this Gospel passage.

In Fleta also, in the chapter on "Of Persons," a "person" is identified as existing from the time of conception, to wit:

> A man may be called freeborn who immediately at birth is free, being the offspring either of freeborn parents or a freedman and freedwoman or of one who is freed and another freedborn, or he may be born of the marriage of a neif and a freeman or of the union of an unmarried freewoman and a bondsman. In the latter case it suffices if the mother is free and unmarried when the offspring is conceived or at any time before birth if, after conception, the pair are united in marriage, because the misfortune of the mother should not injure him who is in her womb.[97]

Comment:

Fleta also cited Bracton as authority, and pronounces the same precedent that a person exists from the time of conception (A "thing" or nonperson could have no rights). But, as mentioned earlier, this holding extends to much earlier times than either Bracton, Fleta, or King Henry I (see Comment under King Henry I, ante).

* * * * *

King Henry VIII

King Henry VIII reigned from 1509 to 1547. His acts have to be viewed legally as clearly a violation of the rights guaranteed by the

Magna Charta, and hence are null and void. By acting arbitrarily outside the law by an exercise of sheer power alone, he actually brought on an anarchical situation. As happens with anarchy, tyranny resulted. But because not even the king was empowered to undo the force of common law precedent, the laws governing the prohibition of abortion remained intact. As to canon law existing at this time, Blackstone noted:

> At the dawn of the reformation, in the reign of king Henry VIII, it was enacted in parliament that a review should be had of the canon law; and, till such review should be made, all canons, constitutions, ordinances, and synodals provincial, being then already made, and not repugnant to the law of the land or the king's prerogative, should still be used and executed. And, as no such review has yet been perfected, upon this statute now depends the authority of the canon law in England.[98]

Comment:

Blackstone cited hosts of Catholic Church canons, decrees, decretals, constitutions, extravagantes, national canons, national synodal constitutions (which are ecclesiastical laws), and provincial constitutions, which remained in force because no review by parliament was ever perfected. Some of these canons extended back to the time of Pope Alexander III (1159-1181). As already mentioned, he specifically included, as part of the established common law of England, the 5 books promulgated by Pope Gregory IX (Decretals of Pope Gregory IX), which included Si Aliquis!! (which see, ante).

* * * * *

Pope Sixtus V

In 1588 Pope Sixtus V issued his Bull, *Effraenatam,* expounding extensively on the evil of abortion and contraception, incorporating past Church history condemning the practice. The following excerpts reveal the official uncompromising attitude the Church has always held toward those crimes (and which necessarily become a part of the common law of England):

> ...Therefore for a good reason the Sixth Synod of Constantinople has decreed that persons who give abortive medicine and those who receive and use poisons that kill fetuses are subject to punishment applied to murderers and it was sanctioned by the old Council of Lleida that those that were preoccupied to kill fetuses conceived from adultery or would extinguish them in the wombs of mothers with potions, if afterwards with repentance would recur to the goodness and meekness of the Church, should humbly weep for their sins for the rest of their lives and if they were Clerics, they should not be allowed to recuperate their ministry and they are subject to all Ecclesiastic law's and profane law's grave punishments for those who nefariously plot to kill fetuses in the uterus of childbearing women or try to prevent women from conceiving or try to expel the conceived fetuses from the womb.
>
> 1. ...We are willing to exterminate in our times also this evil as much as a [sic] We can by the strength given to Us by the Lord: All and whosoever men and women of whatever state, degree, order, rank and condition, even Clerics, secular or pertaining to whatever religious Order, of whatever dignity and Ecclesiastic or worldly illustrious preeminence who by themselves or interposed third persons procure abortion of fetus so that it is expelled by means of blows, poisons, medicines, potions, weights, burdens, work and labor imposed on a pregnant woman, and even by other unknown and extremely researched means so that really abortion follows, and even the same pregnant woman, who knowingly did the aforementioned, incur in sanctions and punishments established by divine and human laws and by Canonic Sanctions and Apostolic Constitutions and which the civil and profane law inflicts upon true murderers and assassins who have actually and really committed murder (here We accept the concepts and terms of all these laws and want them to be literally inserted in this document) and by this our perpetually valid Constitution We state and order that the same punishments and laws and Constitutions are to be extended to the aforementioned case.
>
> 4. We want that those who are subject to Ecclesiastical courts and have committed aforementioned crimes be deposed and

degraded by an Ecclesiastical Judge and be handed over to civil court and secular power in order to be punished in the same manner as it is disposed by divine and profane civil laws against laymen truly and really murderers and assassins.

5. In addition We absolutely establish and decree that the same punishments are to be applied to those who give to women sterilizing potions, medicine and poisons in order to impede conception of the fetus and upon those who make and prepare such potions, medicine and poisons and upon those that give such a counsel, as well as on women who knowingly take such sterilizing potions, medicine and poisons.

7. Besides We want that the monstrous gravity of these brutal, cruel, ferocious and inhuman crimes be punished not only by temporal sanctions but also by spiritual censures and for this reason We decree that all persons of whatever state, degree, Order or condition, laymen as well as Clerics, secular as well as religious of whatever Order, as well as secular laywomen or women professed in whatever religious Order, who as principal parties or accomplices in order to commit aforementioned crimes have helped, counseled, shown favor or knowingly given potions and/or whatever kind of medicine, have written private letters, or have given receipts, and/or identification cards or made payments, or by other words and signs have helped or counseled besides the aforementioned sanctions are also "ipso facto", "latae sententiae", automatically excommunicated by Us.[99]

Comment:

Here Pope Sixtus V decrees that abortion or contraception incur automatic excommunication, and that the persons involved in commission of these crimes are to be considered and treated as murderers. Pope Gregory XIV issued Sedes Apostolica in 1591 removing these harsh penalties which stated "where no homicide or no animated fetus is involved, not to punish more strictly than the sacred canons or civil legislation does,"[100] but, by this time, the accepted common law for abortion had already been firmly established as enunciated by Bracton, which was in accordance with the holdings of the Catholic Church. Pope Sixtus V's and Pope Gregory XIV's decrees came after King Henry VIII ("after the time of Popery") had disclaimed Papal authority, and are not cited by Blackstone as part of the common law of England.[101] *However, under the original immutable principles laid down by the common law of England the Catholic Church's rulings on moral law, and as modified from time to time, were to be the controlling foundation for establishing the common law of England.* ***Comprising a part of the common law of England, this established precedent continues in perpetuity.*** *In this particular instance regarding abortion law (the decrees of Pope Sixtus V and Pope Gregory XIV), however, it was of little consequence because the established legal precedent cited by Bracton had become the controlling precedent and required only a modification of the degree of punishment for early versus later term pregnancy abortions to comport with the mediate animation theory. This temporary reversion to the mediate animation position by Pope Gregory XIV was permanently abolished in favor of immediate hominization in 1869 when Pope Pius IX declared the penalty to be excommunication for abortion carried out at any time in pregnancy.*

Pope Sixtus V also comments on St. Jerome's views on abortion in *Effraenatam* as follows:

> Sorcerers and evil magicians says the Lord to Saint Moses, you will not suffer, allow and tolerate to live: because they oppose overly shamefully against God's will and, as St. Jerome says, while nature receives seed, after having received nurtures it, nurtured body distinguishes in members, meanwhile in the narrowness of the uterus the hand of God is always at work who is Creator of both body and soul and who molded, made and wanted this child and meanwhile the goodness of the Potter, that is of God, is impiously and overly despised by these people.[99]

Comment.

St. Jerome's views have often been cited as endorsing St. Aquinas' views on mediate animation. Here we see the Church's official interpretation of his position, and the Pope citing it in sup-

port of his holding for a condemnation of abortion at any stage of pregnancy.

*   *   *   *   *

Sir Edward Coke

Sir Edward Coke (1552-1634) was Chief Justice of Common Pleas in 1606 and of King's Bench in 1613. His *Institutes of the Laws of England (4 Parts)* are a detailed and expansive compilation of the laws of England up to his time. He takes up a discussion of the Magna Charta as well as common laws, statutes, and customs extending to ancient times.

In order to correctly interpret the common law as recorded by Coke it is necessary to understand his view of the controlling force of established principles in the common law. For example, he points out that the Magna Charta was adjudged by the parliament (25 E. 1) to be taken as the common law.[102] By that same act of parliament he comments that any judgments given contrary to any of the points of the great charter, by the justices, or by any other of the king's ministers, etc. shall be undone, and held for naught.[103] At another place, he goes on to say:

> "*It is a principle in law, &c.*" *Principium, quod est quasi primum caput*, from which many cases have their originall or beginning, which is so strong, **as it suffereth no contradiction**; and therefore it is said in our books, that ancient **principles** of the law [a] ought not to be disputed, *Contra negantem principia non est disputandum.*[104] (Emphasis supplied with bold letters).
>
> [a] 11 H.4.9.

Finally, the most important case which he decided as judge, was *Dr. Bonham's Case,* 8 Co. Rep. 107a, 114a C. P. 1610. His ruling confirmed the preeminent position of the common law. Citing many cases in support of his holding, he demonstrated the supremacy of the common law in cases where acts of parliament are against right and reason, repugnant, or impossible to be performed. It had important ramifications later as it related to the common law adopted by America on gaining its independence from England. The following law emanated from his decision:

> And it appears in our books, that in many cases, **the common law will controul acts of parliament**, and sometimes adjudge them to be utterly void: for when an act of parliament is against common right and reason, or repugnant, or impossible to be performed, the common law will controul it, and adjudge such act to be void....(Emphasis supplied).

This case was to play an important role in the founding of America, which will be seen later in this Section.

This controlling force of common law is also expressed by Blackstone, which is worth mentioning at this point, to wit:

> For it is an established rule to abide by former precedents, where the same points come again in litigation; as well to keep the scale of justice even and steady, and not liable to waver with every new judge's opinion; as also because the law in that case being solemnly declared and determined, what before was uncertain, and perhaps indifferent, **is now become a permanent rule, which it is not in the breast of any subsequent judge to alter or vary** from, according to his private sentiments: he being sworn to determine, not according to his own private judgment, but according to the known laws and customs of the land; not delegated to pronounce a new law, but to maintain and expound the old one.[105] (Emphasis added).
>
> Yet, this rule admits of exception, where the [precedent] **be contrary to the divine law**.[105] (Emphasis added).

Importantly, in his *Institutes*, Coke, documents the common law pertaining to abortion:

> If a woman be quick with childe, and by a potion or otherwise killeth it in her wombe; or if a man beat her, whereby the childe dieth in her body, and she is delivered of a dead childe, this is a great misprision, and no murder: but if the childe be born alive,

> and dieth of the potion, battery, or other cause, this is murder: for in law it is accounted a reasonable creature, *in rerum natura,* when it is born alive.[106]

Comment:

In analyzing this holding on abortion, it is necessary to apply the binding common law principles cited by Coke and Blackstone, and the decision in Dr. Bonham's Case. For example, it has to be a law that does not abrogate previous longstanding and ancient common law principles governing abortion. If silent on any point, it must be interpreted that that portion of the law remains undisturbed. Remedies may be applied to an existing law such as the degree of punishment that may be imposed, but the principle in the precedent must remain intact.

Coke is silent on the nature of the criminal act when it takes place prior to the time of quickening. The fact that if the child is born dead constitutes a great misprision as opposed to being considered as murder when the child is born alive, is, of course, within the framework of the common law in being able to assign the degree of punishment depending on criteria felt to be necessary to deter commission of the crime and to adequately punish if it is carried out. This does not disparage the criminality of the act at any stage of pregnancy, however. Also, being silent on the matter of what degree of crime is constituted by administration of poison or carrying out a destructive act at an earlier time in the pregnancy simply means the application of earlier established common law precedents apply. The decision outlined is simply a recapitulation of previous precedents, although often erroneously touted as showing that abortion was no crime if committed prior to quickening. The fact that the criminal act is considered a great misprision as opposed to being a simple misprision places it barely below the level of murder in degree. In accordance with precedents existing up to Coke's time, abortion prior to quickening would have been considered at least as a crime, in most instances, homicide. Coke, being silent on pre-quickening abortion, means the prior precedents governing the crime would still survive. Those holdings which existed just prior to Coke held pre-quickening abortion to be homicide. However, this fact does not mean that, under Coke, a greater punishment was being inflicted for a pre-quickening than a post-quickening abortion. Under the common law, there were two kinds of killing—murder and homicide.[107] *Murder was willful killing committed with malice aforethought. Homicide comprehended petit treason, murder, and manslaughter. Thus, under Coke's enunciated law governing abortion, pre-quickening abortion, for example, could be considered as either a great misprision— the same as for post-quickening abortion (See Comment under the following Hawkins segment)—or as manslaughter, a less severe crime than a great misprision. Either application would be in conformity with the principle controlling remedies to established common law precedents. By assigning the degree of murder to an aborted child who breathed after death, would serve as an additional deterrent to the commission of the crime of abortion because of the added risk inherited in attempting to terminate a life in the womb during the later stages of pregnancy.*

The law defined by Bracton, "If anyone forcibly interfere with a woman's internal organs to produce abortion is liable," remains undisturbed and in force. The problem of provability of a crime would be simpler in these instances as opposed to beatings, or taking potions or poisons to produce an abortion.

Also remaining undisturbed and in force is the law spoken of by Fleta, "He, too, in strictness is a homicide...who has given or accepted poison with the intention of preventing procreation or conception." A similar prohibition by Si Aliquis also remains as binding law. Prohibition of contraception, also having a Biblical basis in Onan's sin of coitus interruptus, finds accord with divine law which is supreme in the formation of the common law.[105]

The matter of how quickening entered into the equation, and its significance will be dealt with in the segment discussing Blackstone.

* * * * *

Views of Non-Catholics on Abortion

Although the circumstances of history resulted in the common law being developed in concert with the moral teachings of the Catholic Church and the relationship England enjoyed with that faith, that religion was not alone in condemning abortion and contraception:

> **John Calvin:**
>
> For the fetus, though enclosed in the womb of its mother, is already a human being, and it is a monstrous crime to rob it of the life which it has not yet begun to enjoy. If it seems more horrible to kill a man in his own house than in a field, because a man's house is his place of most secure refuge, it ought surely to be deemed more atrocious to destroy a fetus in the womb before it has come to light.[108]
>
> **Martin Luther:**
>
> How great, therefore, the wickedness of human nature is! How many girls there are who prevent conception and kill and expel tender fetuses, although procreation is the work of God.[108]
>
> **Benjamin Wadsworth:**
>
> [The Fifth Commandment also refers to] poisoners and so to those who purposely endeavor to destroy the life of a child in the womb, whether the woman herself, or another does it.[109]

* * * * *

Matthew Hale

The Common Law of England was summed up by Matthew Hale whose work, *The History and Analysis of the Common Law of England*, was published in 1713. Hale was a contemporary of Sir Edward Coke and served with him on the King's Bench. In 1671 he was raised to the office of Lord Chief Justice of the King's Court which position he held until 1676 when he retired.

In his book, *The History and Analysis of the Common Law of England,* he discloses the following:

> Now as to Matters Criminal, whether Capital or not, they are determinable by the Common Law, and not otherwise, and in Affirmance of that Law, where the Statutes of *Magna Charta, cap. 29. 5 Ed. 3. cap. 9. 25 Ed. 3. cap. 4. 29 Ed. 2. cap. 3. 27 Ed. 3. cap. 17. 38 Ed. 3. cap. 9. & 40 Ed. 3. cap. 3.*[110]

Comment:

This declaration means the abortion laws having been established by common law precedents must always be tried under those established common law precedents, guaranteed in perpetuity by the force and provisions of the laws of the early Anglo-Saxon kings, the Magna Charta, and other statutes. The, "and not otherwise," portion of the declaration means abortion must be determined exclusively by the common law.

* * * * *

William Hawkins

William Hawkins wrote, *A Treatise of the Pleas of the Crown,* in two books, published in 1724 in which he documents the holding on abortion, thus:

> And it was anciently holden, That the causing of an Abortion by giving a Potion to, or striking, a Woman big with Child, was Murder: But at this Day, it is said to be a great Misprision only, and not Murder, unless the Child be born alive, and die thereof, in which Case it seems clearly to be Murder, notwithstanding some Opinions to the contrary. And in this Respect also, the Common Law seems to be agreeable to the *Mosaical*, which as to the Purpose is thus expressed, *If Men strive and hurt a Woman with Child, so that her Fruit depart from her, and yet no Mischief*

> *follow, he shall be surely punished, according as the Woman's Husband will lay upon him, and he shall pay as the Judges determine; And if any Mischief follow, then thou shalt give Life for Life* (Citing Exodus c. 21. v. 22, 23).[111]

Comment:

Several points are worth noting:

1. Here the distinction is "big with Child" with no mention of "quickening" which Coke had listed as a target point for triggering the assigned penalty. That is like saying, "when the woman is visibly pregnant," which is less distinct in establishing time, and may precede quickening.

2. As in the case of Coke, being silent on abortion being carried out in the pre-quickening or pre-big stage does not indicate no penalty was applied for abortion at that period, and everything said under Coke can be repeated here as to its portent and significance. Silence on the matter, triggers the precedents already established under the common law for pre-quickening or early abortions.

3. To clarify the analogy between the Mosaical law and the Coke/Hawkins law on abortion, it is important to know that a direct translation from the Hebrew of the cited scriptural Exodus text reveals that the fetus held, throughout pregnancy, a status equal to that of the mother, and that the penalty for injury imposed likely related solely to the child, and not to the mother.[112] *It must be assumed that Hawkins was aware of this correct translation, but, in any event, it was his clearly stated intent to show that the common law on abortion was meant to comply with the Mosaical. To further cement the point that this was the interpretation to be placed on the revised penalty for late term abortion, he cites Bracton as a controlling authority who had taught that the penalty was the same for early and late gestational abortion. However, he also cites Coke demonstrating that the three rulings (Hawkins, Bracton, and Coke) nicely harmonized by assigning the interpretation herein given in the Comment*

Sections of Coke and Hawkins.

4. Hawkins discloses that the common law as stated is to be agreeable with the Mosaical. As already pointed out an accurate translation of the Mosaical text in question states that the fetus held, ***throughout pregnancy****, a status equal to that of the mother, and that the penalty for injury imposed likely related solely to the child. As related by Hawkins, the Mosaical says that if the child is expelled and lives (no Mischief) there is a reduced penalty, but if Mischief (a spectrum of demonstrable physical harms ranging from simple injury to mortal injury and death.*[113]*) follows, then the offender must pay Life for Life if it results in expelling a dead conceptus. The former happening (demonstration of no Mischief) could occur with a normal delivery,* ***but the latter happening (Mischief) could occur at any stage of pregnancy!!*** *Then why did Hawkins limit the triggering event to "big"? Probably, because later stage pregnancies had historically carried a more severe penalty than early stages he was now saying, "if death occurs in the later stages of pregnancy, the punishment is now no more severe than if it occurs at an early stage of pregnancy," but in reducing the punishment for late stage pregnancies it was still going to be considered a heinous crime at any stage of gestation, namely, a great misprision. He cites Coke in his interpretation that the common law is agreeable to the Mosaical, and Coke is very emphatic in stating the common law as he expounds it is to be in agreement with what Bracton and Fleta have recorded and that "...****it is grounded upon the law of God,****" to wit:*

> And so was the law holden in Bractons time, *Si aliquis qui mulierem praegnantem percusserit, vel ei venenum dederit, per quod fecerit abortivum, si puerperium jam formatum, fuerit; et maxime si fuerit animatum, facit homicidium.* (*which translates:* ***If [there is] anyone who shall have struck a pregnant woman, or shall have given her poison, through which he shall have made an abortion/miscarriage, if the thing-born was already formed, [and] especially if it was quickened/ensouled, he commits a murder.***). And herewith agreeth Fleta: and herein the law is grounded upon the law of God....[106]

Thus, Hawkins, Coke, Fleta, and Bracton are of one mind in holding that the law of abortion as recited by Bracton and Fleta represents the exact common law precedent governing abortion, and that it constitutes murder (changed in Coke's time to a great misprision.). Later, Blackstone, too, recited the precise passage and language of Bracton that was cited by Coke in acknowledging that it constituted the ancient common law precedent for abortion.[114] ***Because a common law precedent cannot be contrary to the divine law***[105] ***and the Mosaical forbids abortion at any stage of gestation, and these authorities all hold to that teaching, abortion is established as a crime at any stage of gestation under the common law.***

It is established, then, with certainty that Blackstone, Hawkins, Coke, Fleta, and Bracton were in complete accord that the common law governing abortion was precisely as set forth by Bracton and Fleta. To hold, therefore, as some have done, that the law changed under Coke and Blackstone is without merit. The fact all those authorities agree the law is grounded upon the law of God confirms the force played by the Catholic Church in providing the foundation for those common laws. Therefore, sure reliance can be placed on the precedents having their origin in the precedents set by Archbishops Theodore and Ecgbert, as well as the synods, councils, Decretals of Pope Gregory IX, canons, and other ecclesiastical laws thereby introduced by the Catholic Church. As pointed out by Scott,[115] *the difficulty in deciding whether a dead fetus which accompanied an early abortion was dead as a result of the blows administered, or from an unsuccessful pregnancy, led to the setting of a less severe penalty for an early abortion, even though the crime under the Mosaical law extended to all stages of gestation.*

* * * * *

Sir William Blackstone

Sir William Blackstone was Solicitor General to the Queen of England and assumed a position as justice of the Court of Common Pleas in 1770. He wrote *Commentaries on the Laws of England*

(1765-69). Considered the most important legal treatise ever written in the English language, it was the dominant lawbook in England and America in the century after its publication, and played a unique role in the development of America's fledgling legal system.[116] In his *Commentaries* he correctly discloses the law regarding abortion under the section of murder:

> To kill a child in it's mother's womb, is now no murder, but a great misprision: but if the child be born alive, and dieth by reason of the potion or bruises it received in the womb, it is murder in such as administred or gave them. (citing 3 Inst. 50, and 1 Hawk. P. C. 80.).[117]

Comment:

It is noteworthy that, in citing Coke and Hawkins as authority for this declaration of the law, quickening was not considered by Blackstone (and neither did Hawkins) as an event for declaring the crime of abortion to be a great misprision. As stated by Blackstone, it is a great misprision for the act to be carried out at any time during the gestational period. This gives uniformity in assigning the degree of seriousness to the crime over the entire period of pregnancy, eliminating a time criterion. As quoted, it means that the crime is now a great misprision when carried out at any time during pregnancy rather than just marking from the time of quickening, and the rationality of this interpretation has been discussed under the Coke and, particularly, the Hawkins segments.

The seriousness of the crime can be seen from the definition disclosed by Blackstone as to what constitutes a misprision:

> Misprisions (a term derived from the old French, *mespris*, a neglect or contempt) are, in the acceptation of our law, generally understood to be all such high offenses as are under the degree of capital, but nearly bordering thereon: and it is said, that a misprision is contained in every treason and felony whatsoever....[118]

Not only is a misprision considered as ***nearly bordering on*** *a* *capital crime, but a <u>great</u> misprision narrows the difference even further.*

Proof that the common law governing abortion had not changed down through centuries can be appreciated by the significance of Blackstone citing Hawkins for governing law covering abortion. On reviewing Hawkins, as previously noted, he cites not only Bracton and Coke as authority for his holding, but to be noted, Hawkins states, "And in this Respect also, the Common Law seems to be agreeable to the Mosaical,...." then reciting and citing Exodus 21:22-23 (See discussion, ante, under Hawkins' segment regarding the significance of this holding). This citation of Hawkins by Blackstone means the same precedent continues that was established more than 1000 years earlier by the Penitentials of Archbishops Theodore and Ecgbert, and almost 1500 years earlier by the council of Ancyra (which cited an even earlier ruling supporting its position). ***Thus, the common law governing abortion had not changed from the earliest recorded time of Anglo-Saxon law to the time of Blackstone, and it means that abortion was recognized as a crime from the moment of conception through all stages of pregnancy.***

In completing the discussion of Blackstone's teaching on the law of abortion it is necessary to give attention to another statement he made and which is so often erroneously quoted as providing the basis for alleging that the crime of abortion only extends to the time of quickening. That misleading statement found in his Commentaries is as follows:

> Life is the immediate gift of God, a right inherent by nature in every individual; and it begins in contemplation of law as soon as an infant is able to stir in the mother's womb.[114]

He cites no authority for this statement, and it seems he was improperly confusing "quickening" and "life." ***There can be no quickening without antecedent life.*** *Blackstone acknowledges that abortion is a crime, but that its gravity has been reduced from homicide or manslaughter to a* ***<u>very heinous</u>*** *misdemeanor [great misprision]. In support of his statement of the ancient holding on abortion, Blackstone correctly cites Bracton as authority. Bracton pointed out* *that the crime of abortion was homicide whether the fetus was formed* ***or*** *quickened. In his section dealing with the crime of abortion, Blackstone cited both Coke and Hawkins as precedent and neither taught that life began at quickening. In fact, Hawkins, as has been explained, ante, in his teaching on the law of abortion, cited Exodus 21:22-23 which, as explained, constituted a very early basis for adoption of the principle under the common law condemning abortion at any stage of gestation. Furthermore, in citing Coke, Blackstone had to be also aware of the former citing Fleta. Fleta not only found the penalty for abortion to be in accord with the traditional common law holdings, but, in addition, even went so far as to call the prevention of conception a homicide. Blackstone, himself, acknowledged that roots in the establishment of common law found their foundation in the ancient laws and specifically referenced the laws of Kings Edgar, Edward the Confessor, and Alfred stating, "These are the laws, that so vigorously withstood the repeated attacks of the civil law...." and "These, in short, are the laws which gave rise and original to that collection of maxims and customs, which is now known by the name of the common law."*[119] *Yet, up until this statement of Blackstone, that life begins in contemplation of the law as soon as an infant is able to stir in the mother's womb, there had been no deviation from the holding under common law that abortion was a crime when carried out at any stage of pregnancy, and constituted "killing" which necessarily required that "life" be in existence in order to define that offense. Blackstone cites no precedent for holding that life begins at the time the infant stirs in the mother's womb, only stating it is so adjudged "in contemplation of law." All common law precedents provided only that the woman be pregnant to trigger the crime of abortion; only the penalty was varied according to the stage of gestation. Many have been disconcerted by Blackstone's comment, "Life...begins in contemplation of law...." Obviously, the law cannot define the time of life as beginning at any other time than the time when it begins, and existing precedents cannot be violated. See under* ***Archbishop Ecgbert of York****, ante, establishing the immutable common law precedents that both "personhood" and "life" have their inception at the time of human conception. Even if, arguendo,* *one were to assign a literal interpretation to Blackstone's statement that life begins in contemplation of law as soon as an infant is able to stir in the mother's womb (or quickening), abortions carried out at any time before that period back to conception would still be prohibited and judged as acts of homicide under the common law precedent established by Archbishop Ecgbert in the footnote to his Law Code 30, and also held by Bracton—to which Fleta, Coke, Hawkins, and Blackstone subscribed.*

Blackstone made his revelation in the Chapter of his Commentaries concerned with "Of the Rights of Persons." Bracton and Fleta had both recited in their respective Chapters on "Of Persons" that an individual becomes a person at the time of conception, and those precedents established binding common law, of which Blackstone had to be aware.

Blackstone, in his chapter dealing with the assignment of criteria governing abortion under the common law, makes the correct disclosure, citing Coke and Hawkins, but then, under his chapter, "Of the Rights of Persons," he introduces a conclusion which, on the surface, might seem contrary to the explicit common law precedents held since the earliest immutable common law precedents were established. His unsupported statement that "Life...begins in contemplation of law...." was undoubtedly drawn in reference to Coke's criterion, "...for in law it is <u>*accounted*</u> *a reasonable creature,...when it is born alive," which, in turn, was tied to the quickening factor. (Emphasis added). This error on the part of Blackstone is brought to light by studying Scott's description of how quickening entered into the abortion equation under Coke, and which did not, when carefully analyzed, deviate from established common law governing abortion, but only provided a basis for meeting evidentiary requirements in order to provide a basis for prosecuting the crime:*

> Indeed, Coke justifies his prescription of homicide law, citing Bracton—to the effect that it is homicide to kill a child in utero who is formed, and even more certainly so to kill an animated child—and Genesis 9:6, indicating that homicide law rightly applies against those who kill humans (here, humans in utero).[120]

Scott then goes on to say:

> ...Coke was merely summarizing and harmonizing the common law authorities with canon law of homicide, to give effect to the whole state of the law.
>
> ...during the formative period of the common law [under secular court jurisdiction], birth alive was the only legally conclusive evidence that one's issue was reasonable, i.e., that it belonged to the human species, the only species of reasonable, created beings in existence in rerum natura. The courts' problem with human victims in utero was that they were unseen. This presented the evidentiary problem of how to establish, to the degree required in a homicide case, when and whether they were human and when they were alive or dead. Stillbirth would prove merely human form; birth alive would demonstrate both functioning human form and "in rerum natura" status at the time of birth.
>
> By extrapolating back to the point of quickening—the time at which the child would, if ever, attain living humanness—the court could conclusively prove that the child was a living human at the time of the post-quickening, abortive act. In other words, if the child were found to be living and human at birth, it would be **retroactively "accounted"** a reasonable creature in rerum natura during the post-quickening period in utero. **In this light, Coke's (correct) statement of birth alive/in rerum natura doctrine in terms of retroactive attribution to the fetus in utero, was simply a specific explication of the function of the evidentiary strictures the common law had come to apply in tandem with the quickening criterion.** (Emphasis added).
>
> Given this context, Coke's statement of the born alive rule, "in law [the child] is accounted a reasonable creature, in rerum natura, when it is born alive" may be seen to address both the actus reus and causation. It addresses the material question of the actus reus: by assessing the child's status at birth the defendant's act could be established as directed against a living human being. Coke's statement addresses causation in that this proof of post-quickening life would establish that the child was alive at the time of the act and so (as in standard conditional relevance doctrine) thereafter allow the introduction of evidence of causation: marks on the body, medical testimony, etc.[121]

The importance placed by the early courts on being able to ascertain whether the in utero conceptus was a human being was based on misconceptions the people of that age held regarding options they felt existed. Horne, writing circa 1300 A. D., and recorded by Scott, expresses this concern:

> Of infants killed ye are to distinguish, whether they be killed in their mothers womb or after their births; in the first case it is not adjudged murder; for that none can be adjudged an infant until he has been seen in the world, so that it be known whether he is a monster or no.[122]

Scott had pointed out that this belief arose in Ancient times and accounted for some of the erroneous views developed by Aristotle when it was believed the fetus in early stages led the life of an animal, a sentient creature without reason. As he remarks:

> [The fetus was considered] not a specifically human animal, but a generic animal stage, as it was thought that a pregnant woman could give birth to an unreasoning creature such as an animal or a (mythical) monster—a belief possibly based on the birth of severely retarded or malformed children and of medical monsters.[123]

Thus, Blackstone's statement that "Life...begins in contemplation of law as soon as an infant is able to stir in the mother's womb" carries no weight except for use as an arbitrary definable time that may be used to render a criminal act provable, nothing more.

It is to be noted that all these aforesaid problems arose in cases in which the abortions were caused by battery or potions. Cause and effect would more readily be established for abortions carried out by forcibly interfering with a woman's internal organs even in the earliest stages of pregnancy. The legal strictures with which the early English courts were confronted no longer exist. Legends, then existing, that human pregnancies could consist of animals or monsters are no longer credible. Blood and urine tests can establish the existence of a human pregnancy within a few days, and ultrasonography can identify the living in utero human being with certainty within a few weeks. Histologic examination of products of conception can establish with certainty the destruction of a human embryo in cases in which abortion has been carried out by curetting or expelling the conceptus from the uterus.

Thus, the superfluous embellishments that were placed on the common law (while retaining its immutable principle) and confounding early courts are now irrelevant, and no longer applicable, so that the original and always existing common laws on abortion, in their pristine form, now continue as the governing common laws on abortion. These include, as described in the discussion of Coke, ante, the following laws, reported by Bracton and Fleta, which still remain undisturbed:

1. The law cited by Bracton, which read, "If anyone forcibly interfere with a woman's internal organs to produce abortion is liable."

2. The law reported by Fleta, "He, too, in strictness is a homicide...who has given or accepted poison with the intention of preventing procreation or conception."

And such was the common law regarding abortion existing at the time of the founding of the United States of America, in England until 1967,[124] and in the United States until 1973.

Abortion in Britain, until 1967, was permitted only to save a mother's life.[124] The several states in America fashioned laws based on the controlling precedents which our nation had adopted from the common law of England at the time of its formation. The English laws of 1967 and the Roe v. Wade ruling in the United States in 1973, of course, are null and void inasmuch as they both violate controlling common law, which cannot be extinguished statutorily.

THE FOUNDATION OF AMERICA

The spark, which ignited the decision of the colonies to seek independence, was lit when England was accused by the colonies of departing from controlling common law as it was being applied to them. As previously noted, *Dr. Bonham's Case*, decided by Coke, pronounced the supremacy of the common law over statutory law. In that case, he declared the [common] law was above the Parliament as well as above the King.[125] James Otis in Boston used *Dr. Bonham's Case* to declare the illegality of writs of assistance, stating, "As to Acts of Parliament. An Act against the Constitution is void: An Act against natural Equity is void: and if an Act of Parliament should be made in the very words of the petition, it would be void. The...Courts must pass such Acts into disuse." The Otis Argument in *Lechmere's Case* has been characterized as the opening gun of the controversy leading to the Revolution. John Adams felt this declaration by Otis breathed new life into this nation stating, "Then and there the child Independence was born." It was felt to have given birth to American constitutional law. Patrick Henry, went further, also using *Dr. Bonham's Case* to argue for the annulment of the disallowance by Britain of the Two-penny Act passed by the Virginia Assembly in 1758. It was claimed that Britain was acting in an unconstitutional manner with regard to the colonies. These cases became known as the Otis-Henry doctrine. **The Otis-Henry doctrine was a necessary foundation both for the legal theory underlying the Revolution and the Constitutions and Bills of Rights which it produced.**[126] The supremacy and permanency of common law precedents became recognized and firmly entrenched in development of the law of the newly formed country of America.

It is now worthwhile to review the fundamental principles on which our country was founded and how they prohibited abortion.

First and foremost, are the principles set forth in the Declaration of Independence which declared to the world the justification for our separating from England, and the principles providing that right. At the outset it is important to note that although Thomas Jefferson was the author of that document, he emphasized that he composed it to express the views of the people at large in the colonies, not just his own private views. In his own words:

> This was the object of the Declaration of Independence. Not to find out new principles, or new arguments, never before thought of, not merely to say things which had never been said before; but to place before mankind the common sense of the subject, in terms so plain and firm as to command their assent, and to justify ourselves in the independent stand we are compelled to take. Neither aiming at originality of principle or sentiment, nor yet copied from any particular and previous writing, it was intended to be an expression of the American mind, and to give to that expression the proper tone and spirit called for by the occasion. All its authority rests then on the harmonizing sentiments of the day, whether expressed in conversation, in letters, printed essays, or in the elementary books of public right, as Aristotle, Cicero, Locke, Sidney, etc.[127] (Emphasis supplied).

That the document was very carefully thought through and conformed to the assertion of Mr. Jefferson, that it was an expression of the American mind, can be further appreciated knowing that it was reviewed and approved by all the members of the Committee of 5 appointed to compose a Declaration, and by the entire Continental Congress after it had undergone detailed scrutiny and discussion. Each one of 56 members of that body signed the document, constituting unanimous consent. In all, 86 changes were made in the original indicating the careful thought given to each item included in it, and that it represented the views of the people of the nation.

Of course, a very significant portion of that preamble is:

> We hold these truths to be self-evident, that all men are created equal, that they are endowed by their Creator with certain unalienable Rights, that among these are Life, Liberty and the pursuit of Happiness.

Comment:

Of great importance, is a recognition that although the common law of England was founded and developed based on its adoption of and adherence to the dogma, doctrines, and beliefs of the Roman Catholic Church, only 1 of the 56 signers of the Declaration of Independence was Catholic. Religions of many persuasions were represented in that group of signatories. Yet, the moral standards set by that body of common law were recognized by the founders of America as embodying the principles and laws that should form the bedrock of the jurisprudence for the new nation. This was a recognition of the universal rightness and relevance these laws embraced without distinction as to religious affiliation or beliefs.

Here we learn our country was founded on the people's profession of the creed that life is an unalienable right and is endowed by our Creator. It cannot be taken away by claiming an intrusion on privacy, expediency, constitutions, or legal enactments. Unalienable means unalienable—nothing whatsoever can take away that right. Since that right was endowed by our Creator it cannot be taken away by any human intervention or means whatsoever. Life begins at the time of conception as recognized by the most ancient common laws and intellectual energies throughout the ages devoted to addressing the matter. For anyone in public office to try to eradicate or suppress that right violates his oath of office promising to uphold the Constitution, and abets the wrongdoing. For any private citizen or group to try to abrogate that right commits the crime of interfering with the rights of a human being.

SUMMARY

The earliest recorded history in the development of the common law of England dates back to the earliest days of the spread of the Catholic religion to that country in the first few hundred years A. D. That religion formed the basis of the common law developed by the kings of the realm in governing abortion. Abortion has always been condemned by the Catholic Church.

England lived under these laws in communion with the Catholic Church for almost 1500 years. Even after the conquest of William the Conqueror the previously developed common law had carried forward. King Henry VIII broke from allegiance to the Catholic faith during his reign from 1509 to 1547. However, the common law, which the earlier kings of England had accepted, remained in place as the common law of that nation.

The Magna Charta, which was written in 1215, had explicitly guaranteed the rights and recognition of the Catholic faith in perpetuity to England and its heirs. This document was recognized as part of the common law of England, thereby guaranteeing that the laws against abortion, which were based on the teachings of the Catholic Church, would be guaranteed and carried forward forever.

First, Bracton, and later, Coke, in finally formalizing and collecting the common law of England demonstrated that the very earliest controlling precedents regarding abortion had been continuously recognized without change, save for reducing the degree and type of punishment meted out for commission of the crime. But the seriousness of the crime was never disputed, always being considered as murder or as barely less than murder, depending on the circumstances.

All of the common law precedents, continuing uninterruptedly, from the earliest days of the common law of abortion to the time of the founding of America, held that abortion was a crime from the moment of conception and was prohibited through all stages of pregnancy. Contraception under the common law was homicide. Fathers had a vested right in the conceptus. These were the controlling precedents governing abortion that the new nation of America inherited at its foundation when it adopted the common law of England.

The supremacy of the common law (which includes the Magna Charta) was recognized by Coke in *Dr. Bonham's Case* in 1610, who also recited in his *Institutes* that an established principle of a common law precedent is not to be changed (and also so declared by Blackstone). The force with which the American colonies viewed the *Dr. Bonham's Case* holding, and the superior and perpetually binding nature of the English common law, was expressed in the Otis-Henry Doctrine when they cited that case and justified their reason for seeking independence from England because that country was not adhering to established common law precedents. Very importantly, the U.nited States Supreme Court in the early leading case, Munn v. Illinois[128] echoed the holding of *Dr. Bonham's Case* and the principles enunciated by Coke and Blackstone regarding the immutability of principles established in common laws. It viewed as permissible the particular legislation remedying defects in the common law of the case before them because,

> It establishes no new principle in the law, but only gives new effect to an old one.

Thus, abortion was held as a crime by America until the U. S. Supreme Court in 1973 stepped outside established law in its Roe v. Wade decision.

REFERENCES

1. The Forgotten Books of Eden, Platt, R. H., Jr., Ed. (Cleveland: The World Publishing Company, 1927), pp. 98-99.

2. Jer 1:5; cf. Job 10:8-12; Ps 22:10-11.

3. Ps 139:15.

4. Didache 2, 2: Sources Chrétiennes, 248, 148; Paris: 1942-, p. 547 (Cited in Catechism of the Catholic Church, The Wanderer Press, 1994).

5. Kleist, J. A., Ancient Christian Writers (NewYork: Paulist Press, 1948), pp. 5 & 16.

6. Ibid., p. 62.

7. The Lost Books of the Bible (Cleveland, New York City: The World Publishing Company, 1926), pp. 145 & 163.

8. The Ante-Nicene Fathers (Grand Rapids, Mich.: Wm. B. Eerdmans Publishing Company, Vol. IV, 1994), p. 192.

9. The Ante-Nicene Fathers Vol. III, Latin Christianity: Its Founder, Tertullian (Grand Rapids, Mich.: Wm. B. Eerdmans Publishing Company, 1993), p. 25.

10. Bede, Ecclesiastical History of the English People (Clays Ltd., St. Ives plc, England: Penguin Books, 1990), p. 27.

11. Ibid., p. 49.

12. Ibid., p. 362.

13. Ibid., pp. 51 & 362.

14. Robinson, O. F., The Criminal Law of Ancient Rome (Baltimore: The Johns Hopkins University Press, 1996), p. 97.

15. Ancient Laws and Institutes of England, Vol. 1, Printed by command of King William IV, MDCCCXL, p. xiv.

16. Adams, H., Essays in Anglo-Saxon Laws (Boston: Little, Brown, and Company, 1905), p. 22.

17. Ibid., p. 24.

18. Jones, Fr. D. A., The Human Embryo in Christian Tradition: A Historical Note. Citing account delivered by the Lord Bishop of Oxford, Richard Harries in a House of Lords Debate (Hansard Vol. 62 1, No. 16, col. 35.37): http://www.linacre.org/embryo.html.

19. Plucknett, T. F. T., A Concise History of the Common Law (Boston: Little, Brown and Company, 5th Ed., 1956), p. 8.

20. Oakley, T. P., English Penitential Discipline and Anglo-Saxon Law in Their Joint Influence (New York: Columbia University, 1923), pp. 136-196.

21. Ibid., p. 139.

22. Ibid., pp. 142-144.

23. Ibid., pp. 148-149.

24. Hale, M., The History and Analysis of the Common Law of England (Stafford, England: J. Nutt, 1713), p. 65.

25. Catholic Encyclopedia: Common Law: wysiwyg://4/http//www.newadvent.org/cathen.09068a.htm.

26. Modern American Law, Vol. 1; Part II (Chicago: Blackstone Institute, Inc., 1969), p. 48.

27. Bede, Op. cit., pp. 111-112.

28. Ancient Laws and Institutes of England, Op. cit., Vol. 2, p. 10.

29. Ibid., p. 15.

30. Ibid., p. 23.

31. Ibid.

32. Ibid.

33. The Canons of the Council of Ancyra: www.ccel.org/fathers2/NPNF2-14/Npnf2-14-34.htm.

34 The Council of Elvira: www.bu.edu/religion/courses/syllabi/m301/canons.htm.

35. Bede, Op. cit., p. 373.

36. Ancient Laws and Institutes of England, Op. cit., Vol. 1, p. 103.

37. Ibid., p. 37.

38. Ibid., p. 36.
39. Ancient Laws and Institutes of England, Op. cit., Vol. 2, pp. 155 & 157.
40. Ibid., p. 155.
41. Ibid., p. 183.
42. Ibid., p. 211.
43. Modern American Law, Op. cit., Vol. 1, Part II, p. 48.
44. Ancient Laws and Institutes of England, Op. cit., Vol. 1, p. 49.
45. Ibid., pp. 67 & 69.
46. Ibid., p. 61.
47. Ibid., pp. 167 & 169.
48. Ibid., p. 169.
49. Catholic Encyclopedia: Common Law: wysiwyg://4/http://www.newadvent.org/cathen/09068a.htm.
50. Ancient Laws and Institutes of England, Op. cit., Vol. 1, p. 229.
51. Ibid., p. 215.
52. Ibid., p. 245.
53. Ibid., p. 247.
54. Ibid., pp. 249 & 251.
55. Ibid., p. 269.
56. Ibid., p. 263.
57. Ancient Laws and Institutes of England, Op. cit., Vol. 2, p. 269.
58. Ancient Laws and Institutes of England, Op. cit., Vol. 1, p. 305.
59. Ibid., pp. 307 & 309.
60. Ibid., p. 309.
61. Ibid., p. 313.

62. Ibid., p. 315.

63. Ibid., p. 335.

64. Ibid., p. 347.

65. Ibid., p. 349.

66. Ibid., p. 359.

67. Ibid., p. 373.

68. Ibid., p. 375.

69. Ibid., p. 377.

70. Ibid., p. 383.

71. Remfry, P. M., Medieval Kings of England: www.castlewales.com/eng_king.html.

72. History of the Christian Church, Book 4, Chapter 09: www.godrules.net/library/history/history4ch09.htm.

73. Modern American Law, Op. cit., Vol. 1, Part II, Ch. V, p. 51.

74. Ibid., p. 56.

75. Ancient Laws and Institutes of England, Op. cit., Vol. 1, pp. 495-496.

76. Medieval Sourcebook: Charter of Liberties of Henry I, 1100: www.fordham.edu/halsall/source/hcoronation.html.

77. Ancient Laws and Institutes of England, Op. cit., Vol. 1, pp. 573-574.

78. Ibid., p. 574.

79. Ibid., pp. 582-583.

80. Ibid., p. 627.

81. Plucknett, T. F. T., Op. cit., p. 16.

82. Medieval Sourcebook: Constitutions of Clarendon, 1164: www.fordham.edu/halsall/source/cclarendon.html.

83. Medieval Sourcebook: Magna Carta 1215: www.fordham.edu/halsall/source/mcarta.html.

84. Modern American Law, Op. cit., Vol. 1, Part II, Ch. VI, p. 69.

85. Initiatives for Access, Treasures—Magna Carta: http://reactor-core.org/security/magna-carta-1225.html.

86. Personal communication, Dr. Kenneth Pennington, Catholic University of America, Washington, D. C.

87. Noonan, J. T., Jr., The Morality of Abortion. Legal and Historical Perspectives (Cambridge: Harvard University Press, 1970), p. 21.

88. Catholic Encyclopedia, Papal Decretals: wysiwyg://21/http://www.newadvent.org/cathen/04670b.htm.

89. Catholic Encyclopedia, Corpus Juris Canonici: wysiwyg://10/http://www.newadvent.org/cathen/04391a.htm.

90. May, W. E. (Professor of Moral Theology, Catholic University of America), Contraception, Gateway to the Culture of Death, p. 4: www.christendom-awake.org/pages/may/contraception.htm.

91. Blackstone, W., Commentaries on the Laws of England (Chicago: University of Chicago Press, Vol. 1, 1979), p. 82.

92. Bracton on the Laws and Customs of England, Translated by Thorne, S. E. (Cambridge: The Belknap Press of Harvard University Press, Vol. 2, 1968), p. 341.

93. Ibid., p. 408.

94. Ibid., p. 31.

95. Fleta (London: Selden Society, Vol. IV, 1984), p. xii.

96. Fleta (London: Selden Society, Bernard Quaritch, 11 Grafton Street, W., Vol. II, Book I, Ch. 23, 1955), pp. 60-61.

97. Ibid., Ch. 4, p. 14.

98. Blackstone, W., Op. cit., Vol. 1, p. 83.

99. The Apostolic Constitution, “Effraenatam”: www.iteadjmj.com/aborto/eng-prn.html.

100. The Abortion Decision: www.cath4choice.org/Portuguese/cathwomen/abortiondecision.htm.

101. Blackstone, W., Op. cit., Vol. 1, p. 82.

102. Coke, E., Institutes of the Laws of England (Union, N. J.: The Lawbook Exchange, 2002), Part II, Preface, p. iv.

103. Ibid., pp. vi-vii.

104. Coke, E., Op. cit., Part I, L. 3, C. 11, Sect. 648.

105. Blackstone, W., Op. cit., Vol. 1, pp. 69-70.

106. Coke, E., Op. cit., Part III, p. 50.

107. Ibid., pp. 47 & 54.

108. Choose Life: John Calvin & Martin Luther on Abortion: http://incolor.inebraska.com/stuart/calabrt.htm.

109. *An Essay on the Decalogue or Ten Commandments*, Boston, 1719, p. 29—cited in Pro-Life Activist's Encyclopedia, Ch. 42, p. 2, edited by Brian Clowes, American Life League, P. O. Box 1350, Stafford, VA 22555.

110. Hale, M., Op. cit., p. 50.

111. Hawkins, W., A Treatise of the Pleas of the Crown (London: Printed by Assigns of E. Sayer, Esq., for J. Walthoe in the *Middle-Temple-Cloysters*, MDCCXXIV), Book 1, p. 80.

112. Scott, M. S., *Quickening in the Common Law: The Legal Precedent Roe Attempted and Failed to Use*, 1 Mich. L. & Policy Rev. 199, 203 (1996).

113. Ibid., pp. 201-202.

114. Blackstone, W., Op. cit., Vol. 1, p. 125.

115. Scott, M. S., Op. cit., pp. 204-207.

116. Blackstone, W., Op. cit., Vol. 1, Introduction, p. iii.

117. Blackstone, W., Op. cit., Vol. 4, p. 198.

118. Ibid., p. 119.

119. Blackstone, W., Op. cit., Vol. 1, pp. 66-67.

120. Scott, M. S., Op. cit., p. 232.

121. Ibid., pp. 233-234.

122. Ibid., p. 226.

123. Ibid., p. 211.

124. The Oxford History of Christianity, McManner J., ed. (Oxford, England: Oxford University Press, 1993), p. 390.

125. Schwartz, B., The Roots of the Bill of Rights (New York: Chelsea House Publishers, Vol 1., 1971), p. 182.

126. Ibid., pp. 182-198.

127. Letter of Thomas Jefferson to Henry Lee, May 8, 1825.

128. Munn v. Illinois, 94 U. S. 113, 134 (1876).

Section III

An Analysis of Roe v. Wade

AN ANALYSIS OF ROE v. WADE

General

There were many defects in the crafting of the majority decision of the U. S. Supreme Court (hereinafter, The Court) in the Roe v. Wade case.[1] By dissecting their reasoning and scholarship, those various flaws can be advantageously brought to light.

One of the most glaring defects in The Court's review and evaluation of the law of abortion was its omission of reference to the earliest common laws established in the 7th and 8th Centuries by Archbishops Theodore and Ecgbert, the early Catholic Church councils, and Anglo-Saxon kings. The precedents established at that time are indispensable, essential, and perpetually binding, so it is obligatory to include them in determining the scope and control of the common law regarding abortion.

IT WAS A CULPABLE OBLIGATION ON THE PART OF THE COURT TO INCLUDE THOSE PRECEDENTS IN ITS DECISION-MAKING PROCESS. ITS FAILURE TO DO SO WAS A FATAL ERROR ON ITS PART, ERASING ANY VALIDITY FOR THE CONCLUSIONS IT ARRIVED AT. THIS DEFECT ON THE PART OF THE COURT IS OF VITAL IMPORTANCE BECAUSE ANY INTERPRETATIONS OF LATER LAWS CITED BY SUCH AUTHORS AS BRACTON, FLETA, HAWKINS, COKE, OR BLACKSTONE MUST BE INTERPRETED SO AS TO BE IN HARMONY WITH THOSE EARLIER COMMON LAW PRECEDENTS!!!!

TWO ARTICLES BY MEANS, RELIED ON SO HEAVILY BY THE COURT IN ARRIVING AT ITS DECISION, AS WELL AS THE POSITION ADOPTED BY THE COURT, ITSELF, TREATED THE COMMON LAW ON ABORTION AS HAVING ITS BEGINNING IN THE 13TH AND 14TH CENTURIES.[2, 3] THIS DECEPTIVE MANEUVER DESERVES A SCATHING CON-

DEMNATION. THIS TACK ON THE PART OF THE COURT WAS USED TO PROVIDE A FAÇADE FOR EVADING THE FOUNTAINHEAD ESTABLISHING THE COMMMON LAW GOVERNING ABORTION, WHICH WAS "SET IN STONE" DURING THE EARLIEST DAYS OF THE FOUNDING OF THE REALM OF ENGLAND, BEGINNING WITH THE ANGLO-SAXON SOCIETY AND CONTINUING UNINTERRUPTEDLY FOR CENTURIES BEFORE THE DAWN OF THE 13TH AND 14TH CENTURIES.

THIS FAILURE OF THE COURT TO PROPERLY CREDIT THE TIME FOR ESTABLISHMENT OF THE COMMON LAW, AND TO ABIDE BY ITS PRECEDENTS GOVERNING ABORTION, IS INEXCUSABLE INASMUCH AS BLACKSTONE, WHOM IT CITED AND WAS WELL AWARE OF, HAD EXPLICITLY POINTED OUT ALL THE PERTINENT HISTORY PERTAINING TO THE ORIGIN AND DEVELOPMENT OF THE COMMON LAW AND ALL OF WHICH IT WAS COMPRISED.[4]

Because Means was so successful in providing misleading information—and which was given great weight by The Court in crafting its decision—it will be instructive to preliminarily analyze how that author arrived at his conclusions and the errors contained in them. After completing the analysis of the two Means articles, an analysis of the reasoning used by The Court in formulating its erroneous decision will be undertaken. The two Means articles will be discussed separately and identified as "The Means 1968 Article" and "The Means 1971 Article," respectively. Preliminarily, let it be said that both articles are replete with gross errors.

The Means 1968 Article

In the 1968 article by Means, the following observations may be made:

1. The title of his article is, *The Law of New York Concerning Abortion and the Status of the Foetus, 1664-1968: A Case of Cessation of Constitutionality.*

Comment:

Throughout his article he treats the common law as having its beginning in the medieval ages, specifically, 1664 (See, e.g., pp. 500-501). The 1664 date in his title apparently refers to a time when letters patent appointing the Royal Governors for New York were granted by England making them the surrogates of the Bishop of London. The author states (p. 412):

> As we shall see, the fathers of the English common law fixed it [ensoulment] at the moment of "quickening," a phenomenon which occurs at different times in different women....

There were no "fathers of the English common law" inasmuch as the common laws grew out of customs and usages, dating back many centuries before the concept of quickening was highlighted. This error of failing to correctly identify the common law as having its birth in the very earliest days of Christianity results in a corruption of the interpretation of the laws governing abortion. The Court, in following the same path, was clearly led astray by adopting this same errant reasoning. Because of the fundamental importance the authority of the actual common law holds in fashioning a ruling on abortion, it is difficult to assign inadvertence to this judicial error, especially considering the vast resources available to The Court for its research purposes. A proper recognition of the binding common law precedents governing abortion, established in very early Christian times, would have barred The Court from sanctioning abortions. The concept of "quickening" appearing in Bracton's text was long after the common law was founded and fixed in place (cf. Archbishop Theodore's legal opinion § 4 in Section II). It was only later that Coke, (which see, in Section II) when citing quickening, featured it as a time, not to identify the time ensoulment occurred, but simply to fix an event which could be objectively marked with certainty in order to meet the evidentiary requirements necessary to successfully prosecute the crime of abortion. Many of the early court decisions of our nation, which he cites in support of his view that abortion was not a common law crime, carry no precedential weight because such deci-

sions were premised on beliefs which failed to reference the actual common law dating back to the original laws established more than a millennium earlier. ***To further clarify this matter, abortion was a crime (homicide) for killing a human being at any stage of gestation; an additional penalty was incurred if this killing occurred after the human being was believed to have been ensouled.*** *As disclosed in Section II, ensoulment was held to occur at 40 days after conception, a period much earlier than when quickening occurs.*

2. In the first sentence of footnote 13 on page 419, Means makes the following comment:

> In England, the common law was superseded by Lord Ellenborough's Act, 43 Geo. III, ch. 58 (1803).

Comment:

This law did not supersede the common law, but was enacted to give secular force to the common law, thereby providing an indisputable recognition by the State of the severity of the act of abortion. It is not permissible to supersede the common law. Statutory law is valid only insofar as it remedies a defect in, or enhances, the common law without destroying its principle. The Lord Ellenborough's Act, therefore, was a valid statutory enactment, ensuring the perpetuation of the principle governing the common law of abortion while placing the power of the secular government behind the law to ensure its enforcement.

Scott, discusses the error reflected in the opinion of those who hold a view corresponding to that expressed by Means, and makes the following corrective disclosures:[5]

> Lord Ellenborough's Act recognized that, under the common law, abortion had been a cognizable offense. However, because during the mere 160 year tenure[fn147] of common law [under secular] jurisdiction, the courts had not been successful, e.g., in meting out actual punishment, Parliament took it out of the courts' hands.[fn148]

However, in supplanting court-based efforts, **this Act agreed with the common law** that quickened fetuses merit legal protection, with or without live birth, and also—possibly as a result of then-current preformationist views of human development from conception—extended legal protection to all fetal life. (Emphasis supplied). Thus, Lord Ellenborough's Act made post-quickening abortion a capital crime, and pre-quickening abortion lesser felonies.[fn149] In this, the Act continued the trend, originating in Coke's fusion of canon and common law, of expanding the class of legally protected fetuses. Within 18 years of this Act, the first U.S. state anti-abortion statute was enacted, patterned after the Act.[fn150] A wave of state legislation followed, with the majority of U.S. states enacting such statutes during the period of early modern embryology.[fn151] [fns 149, 150, and 151 not excerpted].

> [fn147] I.e., compared to the over 700 year period of nearly exclusive ecclesiastical jurisdiction over this aspect of the criminal law. Abortion had actually been punishable under the ecclesiastical criminal law, under which perpetrators could be arrested by the sheriff on order of the King's judges.
>
> [fn148] The British statute explained the reason as: "no adequate means [had] been provided for the prevention and punishment of such offenses." In other words, it was the court's failure to apply the common law so as to effect retribution and deterrence that was criticized, not the common law's doctrines; the Act merely expanded the class of fetuses already protected under common law doctrine, following Coke's pattern.

3. In the second sentence of footnote 13 on page 119, Means makes the following comment:

> That was the first statute [The Lord Ellenborough's Act of 1803] on the subject [abortion] in any English-speaking country.

Comment:

Wrong! See Section II and refer to the laws passed by King Edgar (959-975) and King Henry I (1100-1135) declaring abortion to be a crime (The penalty was the same as that assigned for homicide under the common law). In addition, numerous Anglo-Saxon and English kings enacted laws promulgating the canons, ordinances, and Penitentials of the Catholic Church, which always considered abortion to be homicide at any stage of pregnancy for killing a human being, although a heavier penalty was levied for killing a human being felt to be ensouled. Incidentally, England was always an English-speaking country!!

4. Another Means statement found (p. 420) is:

> American courts usually convert Coke's "misprision" into "misdemeanor" in articulating the degree of the offense of abortion *after quickening.*

Comment:

This is the type of error that higher courts should have corrected, and which legal scholars should have highlighted. Misprision was identified by Blackstone (which see in Section II) as a ***serious*** *misdemeanor, and he then went on to disclose that even a misprision was only short of a capital offense by a very narrow margin. The common law holding of what constituted a misdemeanor, then, was far graver than the current concept of the significance of that term. Furthermore, Coke labeled abortion as a "**great** misprision," not a "misprision."*

5. Means, because of his misunderstanding of the history of the common law, drew the following erroneous conclusion (p. 428):

> It is undisputed that the woman herself was not indictable for submitting to abortion, or for aborting herself, before quickening.

Comment:

Under the common law, abortion at any stage of pregnancy was homicide (See e.g., Archbishop Ecgbert and Bracton segments and all the others which so held). Homicide is an indictable offense. However, when the secular courts took over the jurisdiction for handling abortion cases it was obligated to and did adopt and continued the common laws previously established by canons or ecclesiastical laws of the Catholic Church, inasmuch as it was impossible to supplant the principles embodied in the common laws already established. The crime of abortion was certainly an indictable offense, but it was very difficult obtaining an indictment because of evidentiary requirements and a failure of the courts to adhere to the common law precedents. Hence, the reason the Lord Ellenborough's Act, discussed, ante, was enacted. Misbehavior on the part of courts failing to fulfill their responsibility did not abrogate the common laws they were charged with supporting.

It is noteworthy, in contradistinction to Means' statement that, "...the woman herself was not indictable for submitting to an abortion, or for aborting herself, before quickening" that, after 1849, no state passed legislation protecting only quickened fetuses, and in the 1850's, states, originally protecting only quickened fetuses, began amending their laws to protect fetal life throughout gestation, and then-current embryology nomenclature began to be used in lieu of "quick."[6] ***As pointed out by Scott, "In all, the first anti-abortion statutes of 47 American jurisdictions prohibited abortion throughout pregnancy. Although 33 of those did not use the quickening distinction—instead preferring a blanket prohibition throughout pregnancy—of those 17 states using such a distinction, 14 explicitly prohibited abortion both before and after quickening...."*** [7]

6. Another misunderstanding of the common law of abortion is expressed in the following statement of Means (p. 428):

> What was the rule at the common law if the woman desiring the abortion was married? Did her consent alone suffice to make the

act noncriminal, or was her spouse's consent necessary as well?

The author has found no decision or dictum that specifically speaks to this question.

Comment:

Of course, he has found no decision or dictum. Under the common law it was a crime to commit abortion and the husband could not give his consent to commit a crime!! Even the woman committed a crime of abortion if she herself consented to the abortion!! Anyone assisting in the commission of the crime of abortion under the common law, as detailed in Section II, was equally guilty of the crime.

7. Means cites cases decided, which he holds, support the view that the husband has no rights with respect to whether a woman may have an abortion (pp. 428-434):

Comment:

The common law is clear that the husband has rights connected to the fetus. This has been shown in Section II. One of the precedents even stems from the time of the Mosaical revealed in Exodus 21:22-25. Any courts which do not uphold those established precedents under the common law, are not making new common law,—that is not possible—nor following existing common law, but are simply failing to carry out their judicial responsibilities and are deserving of censure. ***It is an exercise in futility for Means to cite court holdings violative of established common law to try to prove that abortion under the common law is not illegal. The established common law on abortion is the common law, not how the courts treat it. To get the courts to fulfill their judicial responsibilities in rendering its verdicts on abortion in accordance with the law is one of the major challenges. Means does not cite any common law in support of his position, only errant judicial decisions. This unacceptable practice of misplaced reasoning, so prevalent in the writings of scholars on***

the subject, is one of the reasons this author felt it was important to set out in explicit unrebuttable view the actual laws themselves to dispel such false conceptions. One cannot prove existing common law by citing instances of violation of the common law!!!! Interestingly, in the case of Touriel v. Benveniste, Civil No. 766,790 (Super. Ct., Los Angeles City, Oct. 20, 1961), the California court ruled the abortion was illegal, and found a husband did have a cause of action, his wife's consent notwithstanding, but Means gives little attention to that case.

8. Means then goes on for several pages allegedly showing that crimes against abortion were enacted simply to prevent harm to the pregnant woman and to attempt to eliminate the danger to her life attendant on various measures to cause the abortion (pp. 434-441). He cites one U. S. court case (State v. Murphy, 27 N. J. L. 112, 114-15, Sup. Ct. 1858) ruling that the law was enacted to guard the health and life of the female against the dangers of abortion (pp. 507-509).

Comment:

Had Means done his research more carefully and completely and gone back to the original common law as set out in Section II of this treatise, he would have realized the futility of trying to "make his case" that abortion laws were made to remove a danger to the pregnant woman. The common law was very specific in identifying the killing of the conceptus as ***homicide****. It is never proper or good law to assign a contrary meaning to the explicit language of a law. All the common law would have said if it meant to protect the woman would be to declare the same penalty for* ***the act*** *of attempting to or completing an abortion* ***for the danger it posed to her****, as opposed to calling an abortion, itself, homicide, as it does. As to the New Jersey case, the common law may be enhanced by adding an additional penalty for placing the mother's health and life at risk, which would serve as an additional deterrent to anyone who wanted to serve as an abortionist. It would also provide an additional deterrent, because, even if the fetus survived an attempted abortion, the abortionist could* *still be held liable for any injury incurred by the woman. Likewise, in the New Jersey case (Means' 1968 article on p. 427), In re Vince, 2 N. J. 443, 67 A.2d 141 (1949), the New Jersey Supreme Court held that* ***"...the New Jersey abortion statute does not cover the pregnant woman*** *herself* ***but only the abortionist..."*** *(Emphasis added), and that the woman was still subject to the common law, although that court erroneously (for all the reasons herein recited in Sections II and III) held that abortion prior to quickening was not an indictable offense. The common law always remains in force, holding that it is a crime to kill or attempt to kill a conceptus; abortion or attempted abortion is unlawful at any stage of gestation, and is an indictable offense. Also, see item 9 under The Means 1971 Article.*

9. Means makes the following comment (p. 435):

> The common law protected the life of the foetus from destruction at its mother's will only *after* quickening.

Comment:

Means is ***totally*** *in error. Reference to Section II reveals a host of explicit controlling common law holdings that abortion is a crime at any stage of gestation. This position represents the precedent established under the common law that has continuously remained in force since the laws of abortion were first developed, and still is the existing and governing common law on abortion.*

10. The following statement is made by Means (p. 438):

> A simple return to our ancestral common law, though sentimentally attractive, is out of the question.

Comment:

There is no option but to adhere to "our ancestral common law" because it is perpetually binding. ***By this admission, Means is***

acknowledging that the "ancestral common law" is contrary to what he is contending it to be!!!!

11. The following statement is made by Means (p. 450):

> Lord Ellenborough's Act was the first in the common-law world to make abortion of a willing woman before quickening a crime, and also the first to elevate abortion of a quickened foetus from misprision, high crime, or misdemeanor to felony.

Comment.

Wrong, again!! Means should have reviewed, for example, the laws enacted by King Edgar and King Henry I (which see in Section II). Their laws assigned penalties for abortion of a willing woman that were those which were consigned for the crime of homicide as detailed by Archbishops Theodore and Ecgbert (which see in Section II). Homicide is a felony. Also, see Comment under item 3, ante. This error by Means as well as that in item 3, ante, is related to his fault in misunderstanding the common law including its foundation and development.

12. Means discloses a New York court decision (p. 486) in which the court ruled that pre-quickening abortion was no crime.

Comment:

As commented before, citation of decisions in violation of the common law do not establish what the common law is. The court, in the cited case, acknowledged there may be life in the embryo. Section II clearly details in explicit terms what the common law was and continues to be. Pre-quickening abortions were a crime under the common law, and that is still binding law.

Author's Additional Comments:

It is instructive to point out that, in our present society, military officers are not excused from criminal liability if they carry out orders of higher authorities that are in contravention of fundamental moral standards characterized as crimes against humanity. Abortion falls into that category of crimes.

Perhaps one of the reasons there was confusion in the early courts and legislatures regarding perceived lack of culpability for pre-quickening abortions is because the early common laws dating back to Archbishops Theodore and Ecgbert, and the early Anglo-Saxon Kings were written in the original Anglo-Saxon language and not translated until much later, and, then, when they were translated, it was rendered in Latin. It may be that these original texts, in translated form, were not available in this country during the formative years of America. The early courts and legislatures, then, may have had to rely directly on Coke, Hale, Blackstone and other authorities who presented the law in English. HOWEVER, WITHOUT THE ORIGINAL LAWS TO REFER TO IT IS NOT POSSIBLE TO CORRECTLY INTERPRET COKE AND OTHER LEGAL SCHOLARS OF THE MEDIEVAL AND POST-MEDIEVAL TIMES.

The Means 1971 Article

In his publication in 1971, Means makes the following disclosures:

1. The title of his article is, *The Phoenix of Abortional Freedom: Is a Penumbral or Ninth-Amendment Right About to Arise from the Nineteenth-Century Legislative Ashes of a Fourteenth-Century Common-Law Liberty?*

Comment:

Means telegraphed the goal he sought to justify. However, he also exposes the false premise on which he will try to build his case. There was no Fourteenth-Century Common-Law Liberty to commit abortion. Review of Section II of this treatise discloses that abortion was continuously a crime under the common law for nearly a millennium preceding and including the Fourteenth-Century.

2. Means makes the following statement (p. 336):

> The original contribution of that article [he had reference to his 1968 article] was the revelation of a truth that had been long forgotten: that the sole historically demonstrable legislative purpose behind these statutes was the protection of pregnant women from the danger to their lives imposed by surgical or potional abortion, under medical conditions then obtaining, that was several times as great as the risk to their lives posed by childbirth at term, and that concern for the life of the conceptus was foreign to the secular thinking of the Protestant legislators who passed these laws.

Comment:

The errors of Means are numerous in this single passage. He made no revelation of a **truth** *in his 1968 article that the statutes were for the sole protection of the pregnant woman. It is true he cites statutory law (p. 435 in his 1968 article) that was enacted for the purpose of making a party guilty of murder for killing a pregnant woman by performing or attempting to perform an abortion on her either with her consent or* **against her will** *(Commonwealth v. Parker, 50 Mass., 9 Met., 263, 1845). That is the way it should be. That is not supplanting the common law of abortion, but only enlarging coverage to protect the woman, as would be expected, especially since the act of abortion itself, per se, is unlawful. See Comment under numbered items 8 and 9, supra, in discussing this matter in connection with Means' 1968 article. The other statutory laws he cites were those enacted which provided that anyone who carries out an abortion on a pregnant woman quick with child, even with her consent, is guilty of an indictable offense. The error made by the courts restricting liability to the post-quickening period of gestation is incompatible with the common law as disclosed in Section II, and a misinterpretation made of Coke's and Blackstone's statements on abortion in which they were silent on the penalty for abortion performed in the pre-quickened period which meant the prevailing common law was recognized as being in effect (it could not be otherwise). See under Coke and Blackstone in Section II for a full discussion of this matter. By failing to acquaint himself with the full common law of abortion from its inception, Means allowed himself to be trapped into erroneously interpreting the common law. It was probably for the same reason the early American courts misinterpreted the quickening proviso, namely, probably the unavailability of prime documents. See* **Author's Additional Comment** *at the end of the consideration of Means' 1968 article. A further discussion of the significance of the quickening provision will be covered, infra, in the discussion of the U. S. Supreme Court's treatment of the Roe v. Wade case.*

When the Commonwealth v. Parker court stated that an abortionist is guilty of murder if either a consenting or unwilling pregnant woman dies from the act of abortion, ***"...on the ground that it is an act done without lawful purpose, dangerous to life...."*** *Means seeks refuge by trying to interpret the life of concern to be that of the mother, and that which makes the act unlawful, because he alleges pre-quickening abortion was not unlawful, but that position presupposes the court is acting in violation of the common law which prohibits abortion at any stage of gestation (as has been amply demonstrated throughout this treatise). So, his position is not a tenable position. In fact, the court specifically stated, "The use of violence upon a woman with an intent to procure her miscarriage, without her consent, is an assault highly aggravated by such wicked purpose, and would be indictable* ***at common law****." The part of the court's holding, "...with an intent to procure her miscarriage....," places the crime outside a simple assault against an unwilling woman. Obviously, the court is announcing its intent to apply governing common law and that would be to find the unlawful act one of abortion rather than simply an assault. Furthermore, the court said, "...the party making such an attempt, with or without the consent of the woman, is guilty of the murder of the mother, on the grounds that it is an act done without lawful purpose, dangerous to life...." The common law governing abortion, as thoroughly documented throughout this treatise, was designed to protect the* ***life of the fetus****, the protestations of Means to the contrary notwithstanding. Therefore, it is only proper to assign the motive of the court as honorable and to be deciding in accordance with the common law holding abortion to be a crime at any stage of gestation.*

Another serious error of Means is his vision that the common law had its inception with the secular thinking of the Protestant legislators. As has been shown in Section II of this treatise, the common law had its birth centuries before medieval times and the reign of King Henry VIII, at which time secular courts took over some of the functions previously handled by the ecclesiastical courts. But as was shown in Section II, the common law was established in compliance with the rules and moral standards of the Roman Catholic Church, and by the controlling influence it exercised in the development of those laws. Those common law precedents as pointed out were firm-

ly established and set in place and were adopted at the time of King Henry VIII as perpetually binding common law. Concern for the life of the conceptus was not foreign to the Protestant legislators and it was the common law that was binding, and this was pointed out by post-Henry VIII legal scholars such as Coke, Blackstone, and Hawkins in citing Bracton, Fleta, and Exodus 21:22-23 as controlling law governing abortion. Not surprisingly, Means does not cite any of the laws supposedly passed by those "Protestant legislators." It can hardly be assumed that Means was unaware that the foundation of the common law dated back to the earliest times of the foundation of the English realm because Blackstone explicitly laid out what was covered in the precedents of the common law,[4] *and Means cited his works on several occasions.*

3. Means discloses the following (p. 336):

> The present article makes a different contribution to the history of this subject, probing the true position of abortion at common law, prior to the nineteenth century. It reveals the story, untold now for nearly a century, of the long period during which English and American women enjoyed a common-law liberty to terminate at will an unwanted pregnancy, from the reign of Edward III to that of George III. This common-law liberty endured, in England, from 1327 to 1803; in America, from 1607 to 1830.

Comment:

The trouble with Means' contribution is that his research was truncated at the 14th Century, but, even setting aside that deficiency, he provided an inaccurate exposé, which was actually a deceptively false portrayal of the common law on abortion. *Reference to Section II of this treatise details the common law governing abortion covering the period cited by Means. Coke, Hawkins, and Blackstone living in that period set out the common law governing abortion, and cite Fleta, Bracton, and Exodus 21:22-23 as common law precedents for what they hold. All those authorities, had to (as they did) declare the act to be unlawful, to reconcile their teachings to be in conform-*

ity with the common law precedents governing abortion extending back to the time of the early Catholic Church and Archbishops Theodore and Ecgbert, for a common law principle once established binds forever. Because it was difficult getting convictions under the common law, England passed a statutory law in 1803 to put teeth in the common law. American legislatures rightfully followed suit. Based on documented historical fact there was no liberty of a woman to have an abortion during the period claimed. ***Everyone is always at liberty to commit a crime, but no one has the right to do so.*** *Just because the perpetrator is not prosecuted or convicted for the crime, does not erase the law prohibiting the crime nor its force. His statement, therefore, is blatantly false. Also, refer to the Comment section of the* ***Views of the Early American Courts on Abortion****, later, under the segment,* ***The Decision of The Court.***

4. Means asks, then answers, the question (pp. 336-337):

> Did an Expectant Mother and Her Abortionist Have a Common-Law Liberty of Abortion at Every Stage of Gestation?
>
> Surprisingly enough, the correct answer to this question is "Yes."

After this introduction he enters into an extended discussion on the subject (pp. 337-352). He cites the *Twinslayer's Case* and *The Abortionist's Case* to try to show that those who committed an abortion were not guilty of a crime at any stage of gestation. In the former case, heard in 1327 (Y.B. Mich. 1 Edw. 3, f. 23, pl. 18: 1327), Means claimed the abortionist was exonerated even though one of the aborted twins died and the other died two days after delivery from the wounds it received from blows to the abdomen of the mother. In *The Abortionist's Case* (Y.B. Mich. 22 Edw. 3: K.B. 1348), decided in 1348, the woman was delivered of a stillborn infant after attempted abortion by blows to the abdomen. An indictment was delivered in the *Twinslayer's Case* as well as the *Abortionist's Case*, but in the latter case no arrest was made because it was difficult to know whether the child died of the blows, and the indictment was judged to be invalid. In the former case, the justices were unwilling to adjudge the thing a felony, and the accused was released to mainpernors, with the argument being adjourned sine die [i.e., without setting a time for reassembling]. There is no record of what happened to the accused while in this interlocutory period. Means admits the case of the still-born child in each instance posed a difficulty-of-proof problem as grounds for results issued by the court (p. 344), but felt it did not explain the result in the live birth in the *Twinslayer's Case*. However, in the *Sim's Case* (K.B. 1601), the justices stated that where a child is born living with wounds appearing on the body, and then dies (e.g., as was the case in the *Twinslayer's Case*), the "Batteror shall be arraigned for murder" (p. 343). Despite these facts, Means held that these two cases demonstrate that under the common law abortion was no crime at any stage of gestation (p. 341)!! Coke also cited the cases in the marginal notes of his *Institutes*, therein relating, "And the book in 1 E. 3. was never holden for law. And 3 Ass. p. 2. is but a repetition of that case." [8]

Comment:

In neither of the cases is there any ground for holding that abortion was not a crime at any stage of gestation. In both cases, establishing proof of the battery causing the stillborns was felt to be insurmountable by the court. This did not render the acts legal. That problem is confronted in every case of the criminal law, namely, to secure evidence sufficient to indict or convict. Yet, an indictment was obtained in both the Twinslayer's Case and The Abortionist's Case, although in the latter case the indictment was ruled invalid because of a difficulty-of-proof problem. And Coke's remarks carry the culpability, of the abortionist even further, undoubtedly feeling the common law held, that battery with intent to commit abortion, should have been prosecutable. ***Nevertheless, it is known that, during the 15th and 16th Centuries, intentionally-induced abortion and tortiously-caused miscarriage were both actively prosecuted in the ecclesiastical courts. For example, the Diocesan "Commissary" court of London heard prosecutions of battery-induced miscarriages and a very early, potion-induced abortion, i.e., of an "infantulum" or "little***

infant." And common law continued to actively develop doctrine in this area.[9] *Also, cases dealing with battery pose a more difficult proof-of-cause than if an abortion were carried out by forcibly interfering with a woman's internal organs to produce an abortion, and this common law prohibition of abortion and cited by Bracton also remains in force.*

The practical problem revolving around the proof-of-cause was not that the in utero conceptus was not considered a human being at all stages, but simply that a causative relationship between death and battery was difficult to establish. Scott discusses this point:

> In the next stage of common law development, while all rationally ensouled infants were still viewed as having achieved human status in fact, legal protection was effectively restricted to a fraction of these late-gestational infants. Andrew Horne (c. 1300) reveals the reason for limiting legal protection, in his explanation of the birth-alive evidentiary criterion:
>
> Of infants killed ye are to distinguish, whether they be killed in their mothers womb or after their births; in the first case it is not adjudged murder; for that none can be adjudged an infant until he has been seen in the world, so that it be known whether he is a monster or no.
>
> Thus, Horne records that the common law, for judicial evidentiary purposes, devised a birth-alive rule, under which only those dying after live birth could be found victims of murder.[10]
>
> * * * * *
>
> Horne's emphasis was that the birth alive doctrine excludes from protection under homicide law those who die in utero and does so for evidentiary reasons: they are excluded because it is not known whether they died as animals, monsters, or human, or were already dead. Horne does not state that homicide law protects only those who are fatally injured ex utero.[11]

Today, with pregnancy tests, the knowledge contributed by modern day embryology, and capabilities for scanning and identifying the in utero contents, there are no barriers to identifying the existence of, and protecting the life of, the human being from the moment of conception. Practical problems revolving around proof-of-cause several centuries ago, no longer exist, and certainly the association of destruction of the conceptus by internal manipulation of the female organs (curettage, saline injections, etc.) presents no problem establishing a relationship between the abortional intervention and the cause of death of the conceptus. It is to be remembered also that the common law always remained in effect that it was homicide to cause or participate in an abortion even though proof-of-cause may have, from a practical standpoint, prevented prosecution of offenders.

5. Means unmercifully attacks Coke, and it is worth repeating in order to show by *Comment*, it was Means, not Coke, who warranted scathing criticism. Here is what Means said (pp. 346-347) about Coke:

> Coke had a politico-religious motive for his outrageous attempt to create a new common-law misprision. One of the reasons for his dismissal from the bench had been his stedfast opposition to the jurisdiction, claimed by the Court of High Commission, a new tribunal among the ecclesiastical courts, to fine and imprison laymen for offenses cognizable at canon law. Abortion had always been an offense within the exclusive jurisdiction of the canonical courts. Coke obviously felt strongly about abortion after quickening ("so horrible an offense should not go unpunished"), and in the privacy of his study he faced a painful dilemma.
>
> If he acknowledged the ancient and exclusive ecclesiastical jurisdiction to try and punish laymen for this offense, then, according to his well-known view, those courts could impose only spiritual penalties (*pro salute animae*) for such offense, such as penances and the ultimate threat of excommunication. Coke knew perfectly well that the laypeople of his time, who had seen so much religious upheaval, were no longer cowed, as their medieval ancestors had been, by the threat of purely spiritual penalties. Those who were bent on having abortions after quickening would go ahead and procure them, and thumb their noses at such threats. For Coke to affirm the traditional ecclesiastical

> jurisdiction, within its traditional bounds, over this offense, would, in effect, be to let it "go unpunished."
>
> If Coke had held that the ecclesiastical courts retained jurisdiction of the offense, but that they could impose really effective punishment for it (fine and imprisonment), his concession would have contradicted those numerous writs of prohibition he had issued, whilst still a sitting judge, to ecclesiastical courts where they had been tried to fine and imprison laypeople.
>
> If, on the other hand, he denied the existence of the ecclesiastical jurisdiction over abortion after quickening, such as had always existed, he either had to let it go unpunished, even by spiritual penalties, or assert a common-law jurisdiction over this offense. He chose the latter expedient.

Comment:

What Means is asking Coke to do is ignore the common law and make new law which would be null and void at its inception!! Coke recognized that the principles contained in the precedents established by the common law are immutable and had their beginning with the foundation of the English realm a thousand years before his time. When the secular courts usurped the jurisdiction from the ecclesiastical courts they nonetheless inherited intact the common law precedents that existed and the secular courts were bound to carry those holdings forward. Blackstone explained what the body of common law contained.[4] *Means cannot escape accountability for knowing of what the body of common law was composed, therefore. These parameters were acknowledged and set by the usurper, King Henry VIII, himself. The Magna Charta and the force of the precedents themselves, guaranteed by stare decisis, assured their existence in perpetuity. Thus, Coke acted very scholarly and honestly. As explained in Section II under the discussion of Coke, the law, as he outlined it, did not abrogate nor attempt to abrogate the common law holding that even pre-quickening abortion was a crime. If he were to be sincere, as he seemingly was, in carrying forward the interpretation of the law he had no option but to hold that abortion was a crime dating back to at least the times of Archbishops Theodore and Ecgbert, and, hence, had to recognize the validity of Bracton's declarations of the law because he reflected the law as it had existed from the time of its origin.*

It is important to note that in laying out his case against Coke, Means acknowledges the "ancient and exclusive ecclesiastical jurisdiction" to try and punish offenders. It is under that ecclesiastical jurisdiction that the common law was originated and established. The secular courts were bound to adhere to those precedents. The secular courts were not free to invent new common law, and Means had to be well aware of that fact. ***Thus, Means acknowledges he was aware that the common law had its origin anciently in the ecclesiastical courts, and, yet, he flagrantly broadcasts the basis for finding what he says the common law holds by assigning its inception as beginning in the 13th and 14th centuries.*** *By criticizing Coke for not recognizing the right to have an abortion at any stage of gestation, Means is condemning him for not violating the law!! Coke, in citing Bracton as authority means he was implementing the actual common law with the exception he was reducing the penalty for the crime of abortion to obtain more convictions, which, in turn, was to serve as an added deterrent to the commission of the crime. These changes corresponded to an increase in conviction rates.*[12]

6. Means, in criticizing Coke, makes the following statements (pp. 347-348):

> While most Americans will sympathize with Coke's objective, to exempt laypeople from the innovative compulsory jurisdiction of ecclesiastical courts to fine and imprison them, in this country we have achieved that goal by a better expedient than the only one that was available to Coke as an author, namely, the clause in the first amendment prohibiting an establishment of religion. The expedient he resorted to, while understandable in a country with a Church established by law, is for that very reason not appropriate to our country.

Comment:

The Church established by law that he speaks of was the Anglican Church. The common law had been established under the aegis of the Roman Catholic Church. Regardless of what institution inherits the responsibility to enforce those laws they remain the same unalterable common laws. America does not escape its obligation to follow and enforce those laws even though its secular courts have jurisdiction. America adopted the common law of England and this was done by the Founding Fathers, with a preponderance of them being non-Catholic. It is not clear what point Means is trying to make, because to whatever jurisdiction it flows to enforce the common law, those laws have already been established and must be upheld.

7. Means makes the following claim (p. 352):

> I have covered the English common law history thoroughly for several reasons:
>
> 1. No modern American scholar has shown any awareness of it.

Comment:

WHOOPS!! You missed the first thousand years of the common law!!!! *The gratuitous self-acclaim Means gives himself is uncalled-for and unjustified.* ***By failing to take into consideration the first thousand years of the common law, Means ignored the most important period of its formation. During those first thousand years the abortion laws found their origin, were "set in stone," and provide to this day the definitive common law governing abortion, and binding in their force in perpetuity. By this fundamental and fatal error of Means in ignoring that period of history, ALL his conclusions are corrupted and invalid!!*** *Means erroneously marked the common law as beginning in the 13th and 14th centuries, but by referring to Section II, it will be seen how thoroughly mistaken he was. Once being made aware of and understand-*

ing this fundamental blunder of Means it is easy to envision and identify the various misconceptions he propounded. Those confused by his teachings will be well served to acquaint themselves with the disclosures contained in Section II of this treatise, in order to fill out the thousand year gap Means failed to cover, and to become correctly informed as to the indispensable and most important period in the development of the common law on abortion.

8. Means attempts to discredit Bracton with the following excerpt (pp. 354-355):

> Writing for the court in *Keeler v. Superior Court of Amador County*, Mr. Justice Mosk observed: "There seem to be no reported cases supporting Bracton's view, and it need not further detain us." The sagacity of this insight was confirmed for me by the leading living authority on Bracton, Professor Samuel Thorne, of the Harvard Law School, who remarked, when I drew Justice Mosk's statement to his attention: "When Bracton had cases to support his view, he cited them."

Comment:

What Bracton sets out on the common law of abortion is nothing more than a recitation of the common law as set forth in Section II of this treatise, spanning centuries. It had to be common knowledge because it was "the law of the land." Even without citations, the law he set down was the common law as clearly defined by the disclosures contained in Section II. Thus, even if Bracton did not cite the specific law, it was still the law!! So, it was valuable, and nice of Bracton to summarize it, for those who might find it a tedious or too daunting an undertaking to review the thousand-year history of the formation of the common law of abortion. Finally, it is important to recognize that those were <u>laws</u> in effect, and did not depend on "cases" for their authority. The authority for Bracton's disclosures of the common law of abortion, as he stated it to be, is found in all of the cited common laws on abortion set out in Section II of this treatise antedating Bracton's arrival on the scene. Finally, Blackstone, himself, *states clearly that the authority of the common law does not have to be found in the written record, to wit:*

> However I therefore stile these parts of our law *leges non scriptae,* because their original institution and authority are not set down in writing, as acts of parliament are, but they receive their binding power, and the force of laws, by long and immemorial usage, and by their universal reception throughout the kingdom.[13]

9. Means cites Sir Matthew Hale's case (p. 362) in 1670, sitting at the Assizes at Bury St. Edmunds where he wrote:

> But if a woman be with child, and any gives her a potion to destroy the child within her, and she take it, and it works so strongly, that it kills her, this is murder, for it was not given to cure her of a disease, but unlawfully to destroy the child within her, and therefore he, that gives a potion to this end, must take the hazard, and if it kill the mother, it is murder, and so ruled before me at the assizes at *Bury* in the year 1670.

Comment:

*Of paramount importance is the declaration of Hale that the act was "...**unlawfully to destroy the child within her**...." Means tries to discount this by speculating that it doubtless meant no more than that such a purpose contravened the King's ecclesiastical law.* ***Undoubtedly, it did mean no more than that because that was the law!!***

*The King's ecclesiastical law, indeed, found abortion to be unlawful. And that **was** the common law, which the secular courts were bound to follow for authority. Means seems to have a difficult time acknowledging even explicit declarations upholding the law if the outcome does not fit his views on "the way it ought to be." This case clearly holds that the laws prohibiting abortion were to protect the fetus, not the mother, in contradistinction to what Means held (See item 8 under The Means 1968 Article, ante.).*

10. Means then relates *The Margaret Tinckler's Case* (pp. 363-371) in which the pregnant woman died as the result of an abortion of a bastard child, and was tried to a jury at the Durham Assizes (Crown Side) in 1781. Sir Edward Hyde East discussed the case in 1803, and reported as follows (p. 367):

> Hither also may be referred the case of one who gives medicine to a woman; and that of another who put skewers in her womb, with a view in each case to procure an abortion; whereby the women were killed. Such acts are clearly murder; though the original intent, had it succeeded, would not have been so, but only a great misdemeanor: for the acts were in their nature malicious and deliberate, and necessarily attended with great danger to the person on whom they were practiced.

Comment:

This decision is in accord with Coke and Hale in their treatment of abortion as a crime. However, Means runs a very convoluted course of reasoning without actually being able to overcome the fact this decision is in accord with established common law making abortion a crime. He again expounds the notion that the decision that was handed down was due to an ecclesiastical/secular court jurisdictional situation. Here is his reasoning (p. 370):

> In this case, too, the stroke is given while the victim is in one jurisdiction ("not yet *in rerum natura*", i.e., in the ecclesiastical jurisdiction), while the death occurs after the victim has crossed into another jurisdiction (through live birth, *in rerum natura, i.e.*, in the common-law jurisdiction). Yet the courts that had exclusive jurisdiction of the stroke (the abortifacient act)—the ecclesiastical courts—remain exclusively competent to try the postnatal death as well.

Then he adds,

> From an American constitutional point of view, of course, the proper line of demarcation between the jurisdiction of secular courts and ecclesiastical courts is of far greater significance than it was for Hale in Restoration England. To Hale, both the com-

> mon law and the ecclesiastical law were emanations of the same sovereign. To us, with the first amendment's disestablishment clause, the sphere of jurisdiction properly belonging to the English ecclesiastical courts is relegated to the domain of private conscience.

Here Means exposes his lack of understanding of the common law. Of course, the common law and ecclesiastical law are emanations of the same sovereign. But the common law is that which was established over a thousand year period before King Henry VIII, and became the law of the sovereign and grew out of the ecclesiastical laws. But, once established, a common law precedent is binding on the secular as well as the ecclesiastical jurisdictions. Common law existed long before the reign of King Henry VIII and was accepted by him as binding. The common law did not have its beginning with King Henry VIII. When America adopted the common law of England, it accepted the common law as it existed from the time of the beginning of the English realm. In America, private consciences neither justify nor permit stepping outside the law. Our disestablishment clause did not free the people of America from being required to follow the law. It is difficult to determine whether Means is confused by a misunderstanding of the common law or whether he is trying to obfuscate the issue.

11. Means states (p. 373):

> The statutory abolition of the *ex officio* oath in ecclesiastical trials thus appears to have been the beginning of the right of privacy in regard to abortion thenceforward enjoyed by English and American women—in England for 140 years (1661-1803) and in America for a quarter of a century more. During the late seventeenth, the whole of the eighteenth, and early nineteenth centuries, English and American women were totally free from all restraints, ecclesiastical as well as secular, in regard to the termination of unwanted pregnancies, at any time during gestation.

Comment:

Wishful thinking!! The common law on abortion always has been in effect from its inception and will exist in perpetuity in England and America. There is no right of privacy that insulates against culpability under the common law precedents governing abortion. As a matter of fact, as will be explained later, again, under ***The Decision of the Court*** *segment this would not even be a permissible modification of the common law because it would abrogate the common law governing abortion, and that would violate the guarantee of perpetual force granted the common laws we adopted. There were no restraint-free periods during the times cited as evidenced by the cases cited earlier in this segment dealing with the Means article of 1971. And as the disclosures surrounding the passage of the Lord Ellenborough's Act pointed out, this piece of legislation was enacted because there was laxity in the enforcement of the common law prohibiting abortion. But the law was always there. "Getting away with committing a crime" does not equate with "freedom from restraint."*

The Decision of The Court

Restrictive Criminal Abortion of Recent Vintage

At the outset, The Court pointed out that the restrictive criminal abortion laws in the majority of the states were of relatively recent vintage (p. 129) and then goes on to say,

> Those laws, generally proscribing abortion or its attempt at any time during pregnancy except when necessary to preserve the pregnant woman's life, are not of ancient or even of common-law origin. Instead, they derive from statutory changes effected, for the most part, in the latter half of the 19th century.

Comment:

It is true that the added provision of permitting abortion to save the mother's life was not incorporated in the inherited common laws governing abortion. In the adopted common laws covering abortion there was no provision for sacrificing the unborn to save the life of the mother. Abortion was a crime under common law without any reservations. And any new laws could not abrogate the existing common law because that was immutable. What it appears the courts attempted to do was to remedy or adapt the common law (without destroying its principle) to accommodate modern advances in medicine when an option might now exist which was not possible previously. Otherwise, no option was available under the common law. Anyway, with present medical advances this problem diminishes to the vanishing point. Now, in our present age, the goal can be focused on saving the lives of both the conceptus and the mother.

Ancient Attitudes

Under its heading, *Ancient attitudes*, The Court (p. 130) states:

> These [*Ancient attitudes*] are not capable of precise determination.

Comment:

This position espoused by The Court deserves particularly harsh criticism, because it is so patently false and insupportable. Analyzing the statement of The Court demonstrates how actually distasteful and misleading this allegation is. For example:

As to, "These [Ancient attitudes] are not capable of precise determination," here are some very precise determinations:

> 1. See under the William Hawkins segment, supra, disclosing that as anciently as the Mosaical law, set out in Exodus 21:22-23, abortion was condemned for destruction of the fetus at any stage of pregnancy.
>
> 2. The earliest code of state-laws in existence was the Sumerian Code of Laws of King Hammurabi, inscribed on a stone monument, and dating from circa 2500 BCE contains the following abortion laws: [14]
>
>> Law 209. If a man strike a free-born woman so that she lose her unborn child, he shall pay ten shekels for her loss.
>>
>> Law 211. If a woman of the free class lose her child by a blow, he shall pay five shekels in money.
>>
>> Law 213. If he strike the maid-servant of a man, and she lose her child, he shall pay two shekels in money.
>
> 3. The Hittite Law Code of circa 1650-1500 BCE contains the following abortion laws: [15]

> Law 17. If anyone cause a free woman to miscarry, if it be the tenth month, he shall give ten half-shekels of silver, if it be the fifth month, he shall give five half-shekels of silver.
>
> Law 18. If anyone cause a female slave to miscarry, if it be the tenth month, he shall give five half-shekels of silver.

4. The Law Code of the Assyrians, circa 1075 BCE, provides the following punishments for abortion:[16]

> Law 21. If a man strike the daughter of a man and cause her to drop what is in her, they shall prosecute him, they shall convict him, two talents and thirty *manas* of lead shall he pay, fifty blows they shall inflict on him, one month shall he toil.
>
> Law 50. If a man strike the wife of a man, in her first stage of pregnancy, and cause her to drop that which is in her, it is a crime; two talents of lead he shall pay.
>
> Law 51. If a man strike a harlot and cause her to drop that which is in her, blows for blows they shall lay upon him; he shall make restitution for a life.
>
> Law 52. If a woman of her own accord drop that which is in her, they shall prosecute her, they shall convict her, they shall crucify her, they shall not bury her. If she die from dropping that which is in her, they shall crucify her, they shall not bury her.

Comment:

Thus, it is absurd to state that the ancient attitudes are not capable of precise determination, and the examples cited do not exhaust the multitude of other similar ones, some of which were mentioned in the **ANCIENT LAWS AND CUSTOMS** *segment in Section II of this treatise.* ***But the only relevant ancient attitudes are those embodied in the ancient Anglo-Saxon and English laws and customs, which have been detailed earlier in this work. Not only are***

those the only relevant attitudes but they are the absolute and indispensable source of the common law adopted by England and America. And, surprisingly, they were completely ignored by The Court in formulating their decision on abortion!!!!

It is to be noted that even pre-quickening abortion was explicitly condemned under Law 50 of the Law Code of the Assyrians.

Father's Rights

The Court stated (p.130):

> If abortion was prosecuted in some places, it seems to have been based on a concept of a violation of the father's right to his offspring.

Comment:

The Court is correct in acknowledging early prosecutions were based on a concept of a violation of the father's right to his offspring. But that right was ***in addition to and concurrent with*** *the penalties levied for commission of the crime of abortion depriving the conceptus of* ***its right*** *to life.* ***And what is very sinisterly strange is The Court in its decision, after acknowledging the right of the father to his interest in the conceptus, did not incorporate that right into its decision, and later stripped him of that right entirely—in violation of common law precedent!!*** *The binding common law guaranteeing perpetually the right of the father to an interest in the conceptus has been adequately disclosed in earlier segments of this treatise. For example, refer to Section II and the segments of* ***King Alfred, King Edward*** *(who adopted the laws of King Alfred),* ***King St. Edward the Confessor*** *(confirming the "Liber Judicialis" of King Alfred),* ***King Henry I*** *(who restored the law of King Edward who, in turn, had adopted the laws of King Alfred), and* ***William Hawkins*** *(wherein is discussed the father's rights attached to the conceptus, dating back as far as the time of* ***Exodus 21:22-23****).*

Ancient Religion Did Not Bar Abortion

The Court stated (p. 130):

> Ancient religion did not bar abortion.

Comment:

In contradistinction to what The Court held, ancient religion, of which the Catholic Church is the ***only*** *pertinent one, did forbid this practice from its earliest days. Already discussed is Exodus 21:22-25, which barred abortion from at least the time of Moses.*

But what is indispensable, and of paramount importance to recognize is that the only religion that has to be considered is the Catholic Church because it is from that religion, and that one alone, that the common law derived its foundation. A consideration of what any other religion held is irrelevant in evaluating the common law binding England and America. It was the stated purpose of the early Anglo-Saxon people and kings to elevate themselves to a higher plane of conduct than the world had lived under before, when people and nations worshiped idols and adhered to other pagan practices. They achieved this purpose by premising their laws on the moral teachings of the Catholic Church. For example, under the segment, "Ancient and Early Common Law of England" in Section II of this treatise, refer to the segments covering Archbishop Theodore (including the references to the Councils of Ancyra and Elvira), Archbishop Ecgbert, and Matthew Hale. Additional cases in point are King Edward's laws, King Ethelred's laws 1 and 34 promulgated in 1008, the Council of Enham's ordinance, and the Ecclesiastical laws of King Cnut, all of which declare the laws and ordinances of the Catholic Church are to be followed, the last mentioned barring abortion from its beginning.

Hippocratic Oath

The Court expends three pages (pp.130-132) devoted to a discussion of the Hippocratic Oath, ending by stating:

> It enables us to understand, in historical context, a long-accepted and revered statement of medical ethics.

Comment:

The motivation is elusive as to why The Court expended such an extended discussion on a subject that has no bearing on the lawfulness or unlawfulness of abortion as controlled by the common law adopted by England and America. Neither England nor America adopted its common law based on the ethical standards of Hippocrates, though he condemned abortion.

The Common Law

Comment:

An important preliminary statement is important regarding the approach taken by The Court in addressing the relationship of the common law in defining its bearing on abortion. Its analysis in this regard was so defective that, in and of itself, the conclusions drawn thereafter were completely corrupted, and led to many serious errors of such a degree that their decision, as will be seen, was so fatally flawed, their conclusions ***frequently and seriously contradicted*** *what the common law actually held.*

To begin with, The Court at no point cited the actual common law as it was brought into existence and developed by the early Anglo-Saxon Kings and the Roman Catholic Church. The earliest common law precedents governing abortion are an integral and inseparable part of the body of common law. It is not possible to properly interpret the later common law citations of Bracton, Fleta, Coke, Hawkins, Hale, and Blackstone, for example, without relating their holdings to the underlying immutable common law principles in which they found their derivation. The historical development of the common law has to be followed and considered from its inception up to the current time. The actual early laws have to be reviewed, exactly as they were enacted, in order to study their impact on follow-on applications and have a proper understanding of later laws. ***Section II of this treatise provides a complete coverage of the common law governing abortion, as it has existed continuously and intact from its inception down to the present time.*** *The Court did a fragmented analysis of the common law governing abortion by only picking selected and limited areas of the common law for use in formulating their decision, and then, in addition, made numerous errors in interpreting and quoting them.*

Reading Section II will bring to the readers attention the inaccuracy and inadequacy of The Court's treatment of the common law of abortion. However, it will be worthwhile, to address specific errors The Court made in considering the various elements they selected for use in their decision-making process.

Pre-quickening Abortion not Indictable

At once, The Court jumped to the matter of a consideration of the role played by "quickening" to define the time at which it was unlawful to carry out an abortion (p. 132), stating:

> It is undisputed that at common law, abortion performed *before* "quickening"—the first recognizable movement of the fetus *in utero*, appearing usually from the 16th to the 18th week of pregnancy—was not an indictable offense.

Comment:

This matter has been discussed at length in Section II, whereby the fallacy of the allegation of The Court is firmly established. Archbishop Theodore in the 7th century referred to "killing" the child in *the womb. He further condemns the expelling of the "ones conceived in the womb" without regard to the period of gestation and lists this under his laws concerning MURDERERS/HOMICIDES.* ***Killing, homicide, and murder are indictable offenses.*** *Archbishop Ecgbert in the 8th century in his laws prohibiting abortion states, "they shall be considered homicides [if an abortion is made]* ***before*** *the child is living [ensouled]* ***or*** *afterwards." He* ***explicitly*** *identifies abortion as* ***homicide*** *at any stage of gestation. King Alfred by his law declared abortion as killing of the child within the womb without regard to the period of gestation and the penalty imposed was that for homicide. King Edgar by his laws declared abortion unlawful, prescribing a penalty assigned for homicide, and the unlawful act pertained to any stage of pregnancy. The laws of King Henry I assigned the penalty for homicide for abortion carried out at any stage of pregnancy. Pope Gregory IX declared, in his Decretal, Si Aliquis, that both contraception and abortion (without respect to the stage of pregnancy) constituted* ***homicide****, and* ***Blackstone explicitly mentioned the Decretals of Pope Gregory IX as being part of the common law of England as well as the rulings of the early Roman Catholic Church and the early Anglo-Saxon kings!!!!*** *Bracton (which see)* ***explicitly*** *classified an act of abortion as* ***homicide*** *without regard to the stage of gestation. Except for the punishment imposed, these holdings of the common law governing abortion were never changed. Bracton also held* ***criminal liability*** *was incurred by anyone forcibly interfering with a woman's internal organs to produce an abortion, without regard to the stage of pregnancy. The Court completely ignored that holding of the common law which, again, constituted abortion to be an indictable offense. Fleta even taught (like Pope Gregory IX) that it was the common law that to attempt to artificially prevent conception was* ***homicide****. The Court, likewise, conveniently forgot to cite that criminal and indictable act. And, had the interpretation of The Court been accepted regarding an indictable offense not being triggered by a pre-quickening abortion, they would of necessity have had to hold that the laws published by Coke, Hawkins, Blackstone, and the other authorities they relied on for their holding were invalid (which they could not do), because they would thereby violate the* *doctrine of stare decisis controlling the abortion common law precedent principles which existed over many preceding centuries, and which established that abortion at any stage of gestation is a crime!! Hawkins, Blackstone, and Coke all cite Bracton for authority for their holdings. In addition, Hawkins invokes Exodus 21:22-23 for authority and that source assigns abortion as a criminal act at any stage of pregnancy.* ***The Court's stated conclusion, therefore, is manifestly and provably untrue.*** *The citations the court referred to as authority for their holding that pre-quickening abortion was not an indictable offense, either do not support its position, or hold contrariwise, as was shown by discussions at length of those cites (See Comments for Coke, Hawkins, and Blackstone in Section II of this treatise as well as the Comments under items 5 and 9 of The Means 1968 Article and item 4 of his 1971 Article.).* ***Once again, it is important to emphasize that The Court tried to find support for its position on abortion by ignoring the common law precedents governing that crime, namely, the first thousand years of common law, which furnished the indispensable and perpetually controlling precedents!! And, as mentioned, even the sources they cite as authority often teach contrary to the position The Court alleges they hold, and in others (E.g., Means) there is no defensible authority cited that supports the view espoused!! What can be definitively stated is that the documented controlling common law precedents provably establish that abortion was a crime when carried out at any stage of gestation, and that even contraception was homicide!!!!***

It is true that many of the early American courts held erroneous opinions as to what the common law held but those opinions were largely based on how abortion cases were being prosecuted rather than on how they should have been prosecuted under governing common law. The difficulty obtaining indictments was transposed by some American courts and scholars into an interpretation holding that pre-quickening abortions were not indictable. However, the actual common law, holding otherwise, was well documented, and ***it was the duty of The Court, not to adopt, but to correct the misunderstanding of legal scholars and courts, educating them as to how***

the actual common law governing abortion stood.

Scott, too, criticizes The Court for how they evaded the common law in arriving at their ruling in wiping away the status of abortion as a crime. The following excerpts provide selected points of criticism he directed at them:

> In summary, abortion was a recognized evil in English Society, and an offense originally handled and punished by ecclesiastical authorities, with spiritual penalty and, ultimately, bodily punishment and imprisonment. **The English ecclesiastical courts actively prosecuted tortiously-caused miscarriages as well as intentionally induced abortions. The early English common law position largely accorded, in Fleta's time, with the Canon Law, and ecclesiastic court activity forms part of the common law heritage.**[17] (Emphasis added).
>
> * * * * *
>
> This historical overview of quickening and its waxing and waning wse (sic), suggests that the Roe opinion was at variance with the Anglo-American legal tradition. The Roe court asserted that the common law's relative silence on abortion (though acknowledged as an offense long under ecclesiastical criminal law jurisdiction and as a practice denounced in common law as deserving of punishment) illustrates the general recognition of a right to pre-quickening abortion. This very association displays the Court's ignorance of what quickening meant and why it lost its validity, as well as the Court's blindness to the social and legal significance of the church courts; its disregard of Lord Ellenborough's Act's explicitly addressing a widely recognized evil not successfully remedied by the common law; and its discounting that American statutes continued a policy of protecting fetal life for its own sake.[18]
>
> * * * * *
>
> So, rather than representing the growth of a new leaf where an old one had withered, Roe's right to abortion amounts to a foreign graft on the Anglo-American legal tradition: one imposed from without, rather than justified from within. In other words, the Court's overturning of state anti-abortion laws, when these

> laws' policy basis remained valid and unchanged—that is that all living members of the human species, in utero and ex utero, should be protected in their own right—is at best unprincipled.[19]

Scott noted, as indicated above, that tortiously caused miscarriages as well as induced abortions were indictable offenses under the common law as prosecuted in the ecclesiastical courts, which could even incur bodily punishment and imprisonment. Under the existing common law those cases would include pre-quickening abortions (See all the common law segments in Section II).

It would be gratuitously ingenuous to assign this error to oversight or lapse considering the immense legal research resources available to The Court. One has to be very skeptical of the reasons for The Court never referencing the very laws that set the controlling precedents for abortion, instead of citing erroneous references or misstating holdings of cases. Ignorance of the law provides no excuse for a charged defendant, and The Court, least of all, cannot escape culpability by this deficiency.

Mediate Animation

The Court pointed out that there was little agreement as to the time ensoulment took place, which ordinarily was believed in early times to be between 40 and 80 days after conception. This belief constituted the philosophy of mediate animation. The Court then went on to say (p. 134):

> There was agreement, however, that prior to this point [40 to 80 days after conception] the fetus was to be regarded as part of the mother, and its destruction, therefore, was not homicide.

Comment:

This conclusion of The Court was drawn out of whole cloth. The Court cites no authority for this holding, and there are none, because they would be in violation of established common law precedent. See Section II and review all the various laws and precedents together with the comments, beginning with Archbishops Theodore and Ecgbert, ***all*** *of which held that it constituted* ***homicide*** *to carry out abortion at any stage of pregnancy. Particularly, read Archbishop Ecgbert's law codes and Bracton's teaching. Conveniently, as previously remarked, The Court never referred to the early common law precedents because, obviously, they destroy any attempt to justify abortion at any stage of pregnancy. They also omit reference to Bracton's declaration that to interfere with a woman's internal organs to produce abortion was a crime, and Fleta's holding that even contraception was* ***homicide****.*

Pope Gregory IX, with the promulgation of Si Aliquis, declared both contraception and abortion at any stage of gestation to be ***homicide****, and the Decretals of Pope Gregory IX (which included Si Aliquis) were explicitly proclaimed by Blackstone (See Blackstone segment in Section II) to be a part of the received common law of England. The conceptus cannot be regarded as part of the mother because its genetic code is* ***ALWAYS*** *different from that of the mother (at every stage of gestation). Furthermore, the act of abortion, in the pre-ensoulment stage, would not be considered murder or homicide if the fetus, during this period, were a part of the mother. Yet, as* ***unrebuttably*** *documented on many occasions, the common law assigned the gravity of the crime as murder or homicide at this stage of pregnancy.*

The Court, in its footnote (pp. 133-134), also cited the Augustinian view supporting mediate animation as being included in the Gratian Decretum and made a part of the Code of Canon Law, suggesting binding Church support of that position. However, the Gratian Decretum was simply a compilation of informational material. The Decretum contained many conflicting texts and antinomies.[20] *The Decretum was not a collection of canonical texts, but a general treatise, in which the texts cited are inserted to help in establishing the law.*[21] *The Decretum may be considered simply as work materials.*[20] *Despite the fundamental importance of Gratian's Decretum, in the middle ages and beyond, it was never formally promulgated by the Church.*[22] *The lack of any binding legal force of the Decretum is attested to by the fact that after its publication in 1140 containing the*

Augustinian concept of mediate animation, 1) Pope Gregory IX, promulgated Si Aliquis in 1234, declaring abortion at any stage of pregnancy to be homicide; 2) then, even after Pope Gregory XIII approved the Decretum in 1582,[20] Pope Sixtus V in his Bull, Effraenatam,[23] in 1588 declared that the woman and any assisting her in having an abortion are to be punished as murderers and assassins without regard to the stage of pregnancy; and 3) Pope Pius IX in 1869 held all abortions were punishable as murder regardless of the stage of gestation.[24]

The Court again raised the issue of mediate animation later in its ruling (pp. 160-161):

> The Aristotelian theory of "mediate animation," that held sway throughout the Middle Ages and the Renaissance in Europe, continued to be official Roman Catholic dogma until the 19th century, despite opposition to this "ensoulment" theory from those in the Church who would recognize the existence of life from the moment of conception.

Comment:

This allegation of The Court is patently false. "Mediate animation" has never been declared a dogma of the Roman Catholic Church. Reference to the immediately preceding "Comment" segment documents rulings by Popes that were contrary to the concept of mediate animation. This type of misquote and other inaccurate presentations of positions, characterizes a substantial portion of the Roe v. Wade decision, resulting in a total analysis of the entire case being required in order to correct the many errors or misconceptions rather than being faced with just having to contend with a few key legal points.

Quickening

On the matter of quickening, The Court stated (p. 134):

> The significance of quickening was echoed by later common-law scholars and found its way into the received common law in this country.
>
> Whether abortion of a *quick* fetus was a felony at common law, or even a lesser crime, is still disputed.

Comment:

After acknowledging that the significance of quickening found its way into the common law of this country, in its next breath The Court states it is disputed whether abortion of a quick fetus was a felony at common law.

Of course, abortion of a quickened fetus was a felony under the common law because it was specifically designated a homicide by Bracton, and when re-classified as a great misprision the revised seriousness of the offense can be appreciated by Blackstone's documentation declaring, "a misprision is contained in every treason and felony whatever." (See Blackstone segment in Section II.). And the seriousness of a ***great*** *misprision would rise above that. As a matter of fact, abortion was a felony in even the pre-quickened stage of pregnancy as documented extensively in Section II of this treatise.*

Actually, the crime of abortion prior to Coke had always been considered a homicide (See Section II). Rather than rely on the disputations of others, all The Court should have relied on was the documented centuries-old unbroken chain of common law precedents establishing that abortion was always at least a felony. It was not good legal scholarship for The Court to place more value on the <u>commentaries</u> of the law than on the <u>actual</u> written law itself.

It is worth calling to mind that the reason for reducing the gravity of the offense from murder to a great misprision was not because abortion was viewed as any less serious, but, rather, to ensure greater success in prosecuting such a heinous crime. Scott, in discussing this topic discloses how the downgrading of abortion from homicide to a great misprision by Coke corresponded with an effort during the medieval times to secure a greater number of convictions. The following isolated excerpts from his publication provide the background concerning this modification in the law, to wit:

> Prior to Coke, common law juries were loath to impose capital punishment and would, given no other option, acquit more often than justified. Juries were influenced by "the reputation of the accused, the nature of his offense, and—perhaps most important—the punishment he would incur"; and where community standards (or extortion, bribery, or partisanship) tended against convicting of a capital crime, juries "simply nullified the law of felony." Where the law prescribed death for homicides generally, 14-15th Century juries regularly downgraded the offense, altering the facts to justify imposing their own ideas of simple homicide versus murder or to support a mitigated verdict of pardonable homicide/self-defense. Juries regularly acquitted at least 80% of all simple homicide defendants and 50% of all murder defendants.[25]
>
> * * * * *
>
> Finally, while downgrading still occurred, [after changes had been made to try to limit jury discretion] e.g., from grand to petty larceny, the official division of homicide into (capital) murder and (non-capital) manslaughter was a measure which worked with the jury's tendency toward finding a lesser (included) offense where the greater was felt too severe. These changes corresponded to an increase in conviction rates.[26]
>
> * * * * *
>
> It is probable that Coke's concurrent designation of misprision feticide, a lesser (included) offense to homicide feticide, fits in historical context as a court-based approach to work with the jury's tendency to downgrade. Thus, a greater quantum and quality of evidence of causation would be required to prove a case of homicide, a capital offense, than that required to prove a case of misprision.[27]

Views of Early American Courts on Abortion

The Court depicts the early American courts as being split on whether abortion of an unquickened fetus was criminal or not, and even injects its view that it appears doubtful that abortion was ever firmly established as a common-law crime, making the following assertions (pp. 135-136):

> This is of some importance because while most American courts ruled, in holding or dictum, that abortion of an unquickened fetus was not criminal under their received common law, others followed Coke in stating that abortion of a quick fetus was a "misprision," a term they translated to mean "misdemeanor." That their reliance on Coke on this aspect of the law was uncritical and, apparently in all the reported cases, dictum (due probably to the paucity of common-law prosecutions for post-quickening abortion), makes it now appear doubtful that abortion was ever firmly established as a common-law crime even with respect to the destruction of a quick fetus.

Comment:

For those courts that felt that abortion of an unquickened fetus was not criminal under the common law, it was incumbent on The Court to supply them with the information contained in Section II of this treatise. It is impossible to erase the principle of an established common law precedent and centuries of an unbroken chain of the common law precedents holding that abortion was criminal at any stage of pregnancy, should have been sufficient persuasion for the courts to acknowledge that law "written in stone" over such a long period as binding on them. Under the circumstances, The Court had an obligation to enlighten and correct them rather than being cowed, as it was, into submitting to their errant views.

For those courts who translated "misprision" to mean "misdemeanor," The Court should have explained and made clear to them that a misprision is barely less than capital in seriousness and that a great misprision (the designation set forth by Coke in grading the enormity of the crime of abortion under adopted revised punishment practices) was of even greater gravity. And Blackstone[28] *referred to abortion as a "***very heinous*** misdemeanor," not a "misdemeanor," and this very heinous misdemeanor was a great misprision with the level of seriousness as just described.*

And imagine, The Court relying on dictum rather than written law to formulate their decision on how the law judges abortion!!!! This acceptation of insupportable views leaves an ugly blemish on the dignity and prestige of that august body. There is no need to rely on any source but the written laws extending over centuries, and those laws were explicit and unyielding in declaring abortion to be a crime of the greatest magnitude at any stage of pregnancy. Again, a review of Section II will easily bring this fact to light.

The English Statutory Law

The Court details quite accurately the development of the English statutory law from 1803 to 1967 (pp. 136-138).

Comment:

The English criminal abortion statute, the Lord Ellenborough's Act, was passed in 1803. The circumstances surrounding its enactment are set out by Scott:

> Lord Ellenborough's Act recognized that, under the common law, abortion had been a cognizable offense. However, because during the mere 160 year tenure of common law jurisdiction, the courts had not been successful, e.g., in meting out actual punishment, Parliament took it out of the courts' hands.
>
> However, in supplanting court-based efforts, this Act agreed with the common law that quickened fetuses merit legal protection, with or without live birth, and also—possibly as a result of then-current preformationists views of human development from conception—extended legal protection to all fetal life. Thus, Lord Ellenborough's Act made post-quickening abortion a capi-

> tal crime, and pre-quickening abortion lesser felonies. In this, the Act continued the trend, originating in Coke's fusion of canon and common law, of expanding the class of legally protected fetuses.[29]
>
> * * * * *
>
> The British statute explained the reason as: "no adequate means [had] been provided for the prevention and punishment of such offenses." In other words, it was the court's failure to apply the common law so as to effect retribution and deterrence that was criticized, not the common law's doctrines; the Act merely expanded the class of fetuses already protected under common law doctrine, following Coke's pattern.[30]

Comment:

When Scott refers to the 160 year tenure of the common law he is referring to the time the secular courts began taking jurisdiction from the ecclesiastical courts in abortion cases. The common law had existed from the earliest times of the Anglo-Saxon nation. But the common law precedents established by them over the years were received into the secular courts as common law precedents.

This type of statutory law in enforcing the common law without abrogating the principles of the precedents they pertain to is a valid exercise of authority.

The Court then commented on England's Abortion Act of 1967 (pp. 137-138) permitting abortion for various reasons including "...risk to the life of the pregnant woman, or of injury to the physical or mental health of the pregnant woman or any existing children of her family...," under certain conditions, and that, "account may be taken of the pregnant woman's actual or reasonably foreseeable environment."

Comment:

England's Abortion Act of 1967 is null and void, except for the condition when it is necessary to save the life of the pregnant woman, because it abrogates the perpetually guaranteed protection afforded the unborn offspring, and goes beyond remedying the common law of abortion or enhancing it.

The American Law

The Court outlined the course of development of the American laws dealing with abortion (pp. 138-141) including the following information:

> In this country, the law in effect in all but a few States until mid-19th century was the pre-existing English common law.
>
> Gradually, in the middle and late 19th century the quickening distinction disappeared from the statutory law of most States and the degree of the offense and the penalties were increased.
>
> Even later, the law continued for some time to treat less punitively an abortion procured in early pregnancy.

Comment:

What is important to note is that ***The Court acknowledges the law existing in America covering abortion until the mid-19th century was the pre-existing English common law****, which it was. This means that all the factors cited in Section II of this treatise comprise the common law of abortion in America at its founding. That means that any laws, constitutional provisions, or legislative enactments would have to be subservient to and uphold the common law provisions governing abortion already in existence. And the principle governing common law precedent, carries with it, guarantees that the common laws prohibiting abortion could not be abrogated. The cited statutory changes by the States in the abortion laws would be permissible under those controlling principles because they only enforce compliance with what the abortion common law precedents prescribe.*

The Court acknowledges the jurisdiction the States hold in regulating abortion by its statement, "...the law ***in effect*** *in all but a few States...." This jurisdiction cannot be usurped as the Constitution* *guaranteed that except for matters specifically delegated to the federal government, the states retain the rights. Thus, this is an acknowledgment by The Court that the States held original jurisdiction.* ***Also, see Comment under item 5 of Means 1968 Article, ante.***

The Position of the American Medical Association

The Court delineated the changing stance of the American Medical Association (AMA) towards abortion (pp. 141-144). In 1859 the AMA declared its opposition to abortion, condemning the belief that the child is not alive until after quickening, and in 1871 they recommended it to be declared unlawful and unprofessional for physicians to carry out abortions or induce labor without certain safeguards, always with a view to safety of the child. In 1967 the AMA's stated policy was to oppose induced abortion except for documented medical evidence of a threat to the health or life of the mother or other mitigating circumstances. Finally, in 1970 the AMA adopted resolutions restricting the conditions for which abortions might be performed and to be done in compliance with state law and that no party to the procedure should be required to violate personally held moral principles.

Comment:

The opinions of the medical profession are not controlling in terms of the established common law governing abortions. The precedents are now already "set in stone" in which abortion is a crime when carried out at any time of pregnancy. The medical profession and allied sciences, in the meantime, have accumulated a huge amount of scientific data that incontrovertibly establishes that human life begins at the time of conception, and to this extent lends valuable evidence in confirmation of the common law of abortion finding it unlawful to commit this crime at any stage of gestation. Also, many of the concerns of the past as to saving the life of the mother versus that of the fetus have been reduced almost to the vanishing point by advances made in medicine and now making it possi-

ble to ensure the salvage of both mother and fetus.

The Position of the American Public Health Association

The Court details the policy of the American Public Health Association (APHA) in 1970 (pp. 144-146) adopting standards approving abortions, and suggested discussing contraception and/or sterilization with each abortion patient.

Comment:

The Court cannot look to outside groups to justify changing the common law on abortion. The Court's obligation is to ensure compliance with the common law, and not to follow the violations of it proposed by various groups. One wonders at the purpose in reciting what all these various organizations stand for when their edicts or opinions cannot abrogate immutable established common law. So, citing the policies of the APHA is irrelevant. Furthermore, their recommendation in support of sterilization and contraceptives contravenes common law precedents.

The Position of the American Bar Association

The Court recites the position of the American Bar Association (ABA) and then sets forth the Uniform Abortion Act in full (pp. 146-152). Three reasons were advanced to explain the enactment of criminal laws in the 19th century and to justify their continued existence, to wit:

> It has been argued occasionally that these laws were the product of a Victorian social concern to discourage illicit sexual conduct.
>
> A second reason is concerned with abortion as a medical procedure. Thus, it has been argued that a State's real concern in enacting a criminal abortion law was to protect the pregnant woman, that is, to restrain her from submitting to a procedure

> that placed her life in serious jeopardy.
>
> The third reason is the State's interest—some phrase it in terms of duty—in protecting prenatal life.
>
> It is with these interests, and the weight to be attached to them, that this case is concerned.

Comment:

The actual reason for the enactment of the criminal laws of the 19th century was to enforce the common law prohibiting abortion, necessitated because the courts often refused to do so.

The first "reason" finds no basis in fact. It was not a Victorian social concern since the condemnation of abortion went back to Biblical times, and centuries before the Victorian age. Exodus 21:22-25 condemning abortion reached into ancient Biblical times. The Sumerian pagan law 2000 years BCE condemned abortion (There was no Victorian social concern then!). The Didache of Apostolic times condemned abortion. Illicit sexual conduct carried its separate penalties under the common law, and the penalties were less than for abortion. If the intent of the abortion laws was to discourage such conduct, the penalty for the targeted unacceptable conduct should have been as grave as those assigned for abortion itself. There are no historical grounds for believing that our criminal laws were enacted based on legislative deception. The Court would have to impugn America's entire legislative process if it were to take this tack. That would be unbecoming the highest court of our land, and deserves no countenance.

As to the second "reason," there is no historical basis in the development of the common law for the contention that the criminal laws were enacted to protect the pregnant woman. The woman is the captain of her own ship, and, in America (and in England under their early laws guaranteeing liberty) the State would have no right to control her destiny. That would be tantamount to laws being enacted forcing or coercing a patient to accept the least risky of two medical or surgical options. A patient has the right to choose a more risky procedure if he/she deems the potential benefits warrant taking the added risk. If the product of conception were not a human person, even then, the State would have no constitutional right to enact a law which would proscribe the patient exercising her liberty rights to make her own decision once fully informed as to the risks versus the benefits. Therefore, the enactment of the laws could only be intended to protect in utero life, and this is confirmed by the fact all the common laws declared the premature expulsion or destruction of the conceptus as ***killing*** *or* ***homicide****. It would be ingenuous to believe early legal enactments and court decisions were designed to achieve a purpose other than that stated in the law, itself, in order to deceive society—to successfully get away with it—and be able to perpetuate that lie over many centuries. Immutable controlling common law explicitly states abortion was to prevent killing or murder of an in utero conceptus (only possible if it were alive). The letter of the law must be believed, especially when it has been perpetuated unchanged (except for penalties imposed) over many centuries.* ***How repelling it is to even be presented with the thought that our government would lie about its purpose in enacting a criminal law in order to sneak in a hidden intention different than the stated purpose!!*** *Just the practical elements faced in getting a bill passed into law with the extensive legislative process required and all the hearings and officials involved renders the proposition that the criminal laws passed for any but their stated purposes is ludicrous.* ***Imagine the state of affairs that would exist if our society had to live under conditions in which the law as stated could not be depended on as an accurate interpretation of what it meant!!!!***

The third reason stated by the ABA is a valid one, the actual one, and the one ignored by The Court.

Privacy

The Court states (pp. 152-156):

> This right of privacy, whether it be founded in the Fourteenth Amendment's concept of personal liberty and restrictions upon

> state action, as we feel it is, or, as the District Court determined, in the Ninth Amendment's reservation of rights to the people, is broad enough to encompass a woman's decision whether or not to terminate her pregnancy.

and,

> As noted above, a State may properly assert important interests in safeguarding health, in maintaining medical standards, **and in protecting potential life**. (Emphasis added).

and,

> The privacy right involved, therefore, cannot be said to be absolute.

Comment:

The Court is incorrect in stating that the 14th and 9th Amendments are broad enough to encompass a woman's decision whether or not to terminate her pregnancy. The people of England were guaranteed the right to liberty under the Magna Charta, but it provided no basis for an escape valve to evade the common law prohibitions of abortion.

Right to privacy was never a criterion under the controlling and established common law defining the crime or unlawfulness of abortion. And such a condition cannot be grafted onto that precedent since it would destroy the principle of that precedent and that is not permissible. *The prohibitions against abortion established under the common law, which America adopted, are absolute, and cannot be extinguished by any means whatsoever. The women under the ancient common law had as much right to privacy as the women of our present age, but that "right" was never incorporated into the common law nor constituted a reason to permit abortion, and, therefore, in formulating present day legal rulings privacy rights have no standing as a basis for permitting abortion.*

To invoke such a purported right of privacy would violate the immutability guaranteed to the common law principle prohibiting abortion, because it would not be enhancing the common law principle prohibiting abortion, but abrogating it. And that is not permissible, as previously discussed in detail (See Coke, the Dr. Bonham's Case, Blackstone, and the U. S. Supreme Court decision in Munn v. Illinois, 94 U. S. 113, 134; 1876 covered in Section II of this treatise.).

*The genetic code of the conceptus is **ALWAYS** different from that of the mother which means the conceptus is a distinct and separate person from the mother and not her body part.* ***HENCE, THE CONCEPTUS HAS AS MUCH RIGHT TO PRIVACY AS THE MOTHER!!!!***

Adopting the precept of a right to privacy of the mother as a basis for formulating their ruling, as The Court did, renders the Roe v. Wade decision null and void at its inception!

Person

The Court made the following comments (p. 157-158):

> On the other hand, the appellee conceded on reargument that no case could be cited that holds that a fetus is a person within the meaning of the Fourteenth Amendment.
>
> The Constitution does not define "person" in so many words.

and,

> But in nearly all these instances, [referring to their previous citations of Constitutional inclusions of the word, "person"] the use of the word is such that it has application only postnatally. None indicates, with any assurance, that it has any possible pre-natal application.
>
> All this, together with our observation, *supra*, that throughout the major portion of the 19th century prevailing legal abortion practices were far freer than they are today, persuades us that the word "person" as used in the Fourteenth Amendment, does not include the unborn.

Comment:

The only conclusion that can be drawn from The Court's declaration that "person" does not include the unborn, is that the proper homework has not been done establishing, irrefutably, that "person" does, indeed, include a human being from the moment of conception, or the facts bearing specifically on the subject have been conveniently ignored. At the outset of Section III it was pointed out that one of the glaring deficiencies in The Court's crafting of the Roe v. Wade decision was that it had omitted to review common law precedents established in the earliest Anglo-Saxon period. The requirement to do this is indispensable because those common law precedents are still binding on America. This is such an elementary requisite, and so crucial in establishing how the governing precedents of the law are to be interpreted, it is troublesome why this investigation was not done.

*In Section II under the Bracton, Fleta, and King Henry I segments discussion of this crucial point of when personhood is established was undertaken. Therein it is disclosed that since at least the 7th and 8th centuries a child receives its **status** at the time of conception. **Only a person can receive status**. In addition, Archbishop Theodore in the 7th century condemned the **killing** of a **child** in the womb, and **killing** of the **ones conceived** (used in connection with the word "child", as it had reference to, means "person") in the womb. Killing of a child can only mean killing a person (Incidentally, the word "child," under the common law, is used generically to depict any stage of in utero development of the human conception. For example, the Oxford English Dictionary defines "child", thus: "With reference to state or age. The unborn, or newly born human being, foetus, infant. 'Child' is still the proper term, and retained in phrases as 'with child'"). It is to be noted that Archbishop Theodore included under his section, "Concerning Murderers/Homicides," those who "...expel the ones conceived in the womb...." Homicide or murder can only be applied to the killing of a human being—a person.*

Archbishop Ecgbert in the 8th century, declared as constituting

***homicide**, expelling a "thing-conceived in her womb...." Only killing a "person" can constitute homicide.*

*The above precedents, as they must, always remained the established common law precedent covering abortions. Fleta declared abortion as **homicide**. Bracton declared abortion as constituting **homicide** at any stage of pregnancy. Fleta even stated to prevent conception amounted to **homicide**. It was impossible to change these common law precedents extending over many centuries. The reclassification of abortion as constituting a misprision as opposed to **homicide** could not be interpreted as reducing the status of the conceptus to a nonperson. That was impossible to do because a common law principle once established is a precedent that cannot be changed except for the punishment meted out. The misprision was simply eliminating the capital punishment associated with killing a person. In his "Commentaries," Blackstone correctly includes the penalty of abortion under his section dealing with murder. If he intended to mean that abortion was anything other than the taking of the life of a human person his holding would be null and void, as it would be seeking to accomplish the impossible.*

It is well to review Section I and the entire history of the common law in Section II, which documents extensively that the common law held that the in utero conceptus of a woman was a "person" from the moment of conception.

Finally, The Court made the following observation (pp. 156-157):

> If this suggestion of personhood is established, [that the fetus is a "person"] the appellant's case, of course, collapses, for the fetus' right to life would then be guaranteed specifically by the Amendment.

Comment:

In addition to what has been recited above concerning the historical proof that personhood was recognized legally under the common law as existing from the time of conception, it is important

to also refer to the "Life" section which follows wherein it is shown that science also has proven this fact beyond any doubt.

It can be stated with certainty that Roe v. Wade collapses by force of the criterion The Court established, in view of the fact that personhood of the fetus has been recognized and firmly established under the common law (and by science as per the "Life" segment, infra) as existing from the time of his/her conception.

Even though the common law established personhood and human life in its rulings as beginning at conception, it is well to keep in mind that the binding common law precedents regarding abortion are not based on definitions of life (sidestepped, in any event, by The Court), person, or privacy, so it was irrelevant for The Court to introduce them as an underpinning for the conclusions they drew. ***What the common law holds is that if, at any stage of gestation, a pregnant woman or anyone else uses any means whatsoever to cause an abortion, that constitutes a criminal act.*** *Even to prevent conception (contraception) under the common law is homicide. As Bracton pointed out, if anyone interferes with the internal organs of a woman to produce an abortion (See Section II, ante) that is a crime without distinction of time of the gestation.* ***Protection was guaranteed to the conceptus independent of any constitutional or statutory provisions, and any such enactment would necessarily require they support those guarantees, inasmuch as they would be subservient to guarantees included in the common law.***

Life

The Court made these important statements (p. 159) in its ruling:

> We need not resolve the difficult question of when life begins.
>
> ...the judiciary, at this point in the development of man's knowledge, is not in a position to speculate as to the answer.

Comment:

Faced with the insuperability of overcoming the protection the United States Constitution guaranteed a live human being in utero, The Court simply sidestepped the issue to avoid the one issue which would have provided them the legal right to bring the case before The Court for adjudication. Otherwise, it is a matter for the state courts and the case of Roe v. Wade should not have been heard in The Court once they sidestepped this issue. On the matter of protecting the right to life of a person, The Court would have properly held jurisdiction to ensure the State was fulfilling its obligation to protect that life.

It is indefensible for any average person of mature age living in the 20th and 21st centuries to hold that it is a "difficult question as to when life begins" or to feel there are grounds for having to "speculate" as to the answer. It is a confession unbecoming legal scholarship even at an elementary level.

The definition of human life is given in Section I of this treatise.

The Declaration of Independence states that life is an ***unalienable*** *right endowed by the Creator. We are bound by the provisions of that document, because it is our solemn declaration —our promise to structure our society upholding that guarantee—to mankind that it is one of the bedrock principles on which we relied to justify our separation from England. That is a right so profoundly certain it is rightly described as being self-evident. It is well to keep in mind that the right cannot be abrogated on any basis. It is futile, for example, to claim a right to privacy as justification for stripping one of the right to life. If one could evade this right by a constitutional amendment, a statute, a precedent, invoking of a right of privacy, or any other reason whatsoever, it would not be unalienable.*

There is no "difficult question" involved nor is there any reason to "speculate" about a firmly established fact. Every 9th grade biology student knows that the unit of life is the cell. For the human, that life begins with the formation of the human zygote. In Section I reference was made to Arey's textbook and how a given individual has his/her beginning at the time of the formation of the zygote. It is very clear, therefore, that the statement, "...at this point in the develop-

ment of man's knowledge....," was drawn out of whole cloth since it has been known from time immemorial that life begins at the time of fertilization, and science has now so firmly established that fact it can no longer be disputed. So certain is this fact that it can now be said that modern science has closed the issue.

As stated in Section I, Professor Jerome Lejeune, M.D., Ph.D., an internationally renowned authority in genetics and pioneer in the field documented proof (yes!) by his testimony in court[31] *that the human being takes on his/her identity as a person at the time of the formation of the zygote, just as the embryologists (See reference to Arey earlier in Section I) and learned scholars throughout the ages had recognized and taught since time immemorial. Refer in Section I to the testimony of all the internationally distinguished scientific and medical experts upholding this view, including that of Dr. Lejeune in which he testifies he has never heard a colleague dispute that the embryo is not a piece of property, but rather is a human being. One wonders how all this known documented irrefutable evidence existing for ages escaped the attention of The Court in fashioning its opinion.*

Finally, and conclusively, it is absolutely proven that the product of human conception has life (human life) and is a human being by the following indisputable facts:

1. If the conceptus were an inanimate object it would be expelled spontaneously from the uterus of the woman and could not become implanted.

2. If the conceptus had life, but its genetic code was not that of the Homo sapiens species, it would be recognized by the host as foreign and expelled spontaneously from the uterus of the woman.

3. If the conceptus dies in the womb, it is spontaneously rejected and expelled from the uterus. Human conceptus life is required in order for implantation to take place and persist in its continuing developmental process.

4. It is required that the genetic code of the conceptus be that of a Homo sapiens (i.e., a human being) and have life for it to become implanted in the uterus of the woman.

5. Therefore, the conceptus dating from the time of conception <u>MUST</u> a) have life, and b) be the life of a human being (ergo, a person), for it to become implanted in the womb of a woman.

To use the term, abortion, as The Court does, is an admission of the recognition that there is life in the unborn in the womb because that is the goal of abortion—to destroy a life, and terminate its normal developmental process. To remove an inanimate or foreign living material from the womb is not abortion. Abortion is termination prematurely of developing life in the uterus—in the human uterus, a human life. That's what the word means, "failure to complete what has been begun."[32] *Life is a state of continual change. In the human, the normal progressive stages of development are:*

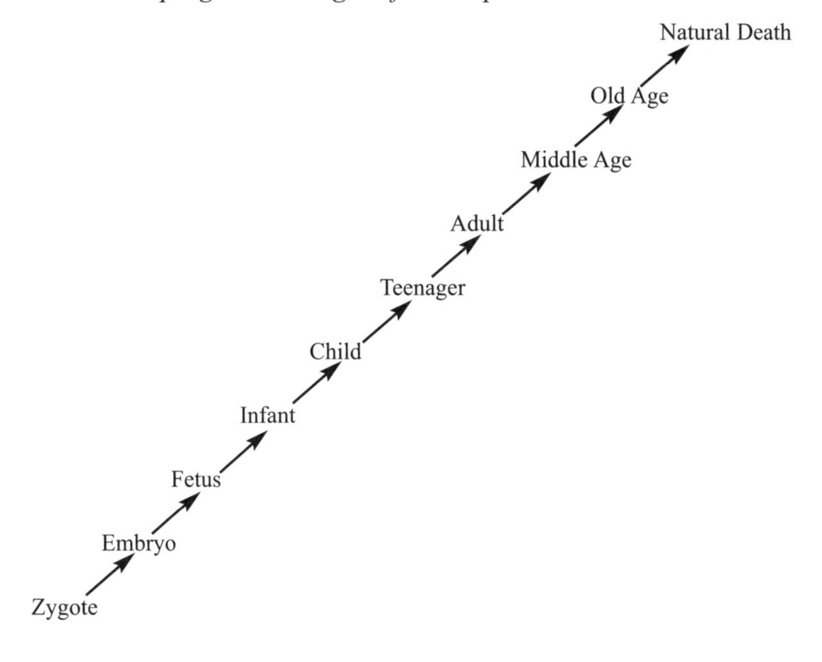

That developmental process constitutes a defining quality of life. Therefore, in all the ancient and recent laws in which the term abortion is used, the reference is to an act recognized as constituting the termination of a human life in the process of development. ***As previously stated, the genetic code of the conceptus is ALWAYS distinct and different from that of the mother, so the conceptus cannot be simply another body part of the mother, but rather is a separate distinct person.***

Thus, since it is proven that human life begins at conception, the conceptus, at all stages of gestation, is entitled to the protection afforded by the 5th and 14th Amendments to the U. S. Constitution.

DISCUSSION

Abortion, under the controlling English common law, which was adopted by America, has been held since its earliest days to be a criminal act. The federal government has no jurisdiction to deal with that crime except that it has only been empowered to uphold the right to life of the unborn as guaranteed by the protection afforded under the 5th and 14th amendments. The holding in Roe v. Wade has been void since its inception for the many reasons hereinbefore documented. The right to try the crime of abortion under established common law devolves, alone, upon the several States.

It is a mistake to seek a constitutional amendment barring abortion or to pass a constitutional amendment or law declaring that personhood exists from the moment of conception because that would acknowledge (falsely) that those rights or designations can be changed at the whim and fancy of the people at any given time. The right to life is unalienable and an endowment from the Creator and cannot be added to nor taken away by man. A person is not defined by statute or constitutional provisions but by the truth of the biological facts, which, likewise, are unchangeable. And, as cited previously, Bracton, Fleta, King Henry I, Archbishop Theodore in the 7th century, and Archbishop Ecgbert in the 8th century have pointed out that under established common law a person exists from the moment of conception. The common law holds, that all that is necessary to commit the crime is for a pregnant woman to procure or have an abortion at any stage of gestation, or for anyone to assist in bringing it about.

No law can be passed by the States permitting abortion because that would be a violation, not only of the unalienable right to life of the in utero human being, but also that individual's constitutional guarantees. It is required that each of the States undertake to prohibit abortion under established common law, enforced with punishments severe enough to discourage the practice and to provide an appropriate penalty for the actual commission of the crime.

Congress has the power and it would be proper for that body to remove appellate jurisdiction from the U.S. Supreme Court in abortion cases, as authorized under Article III, Section 2 of the U. S. Constitution. It would eliminate an area in which the Supreme Court had no right to intrude into in the first place. The only proper role for the U. S. Supreme Court would be to ensure that the States uphold their Constitutional obligation to protect in utero human life.

America, by departing from legal precedents concerning abortion, has reduced our country to the barbaric savagery practiced by uncivilized societies. It has brought dishonor on, and desecrated our country, eliminating the important powerful moral force we had hitherto been able to exert in the world, so necessary to establish a peaceful and successful international society. Throughout the first, almost 200 years of the history of our nation, we were proud of the particular distinction that distinguished our country from other countries of the world by the sacredness we attached to the value of human life. In the 1960's a rebellious spirit took over our country, and was inflamed in 1973 by the Roe v. Wade decision so that the lower courts were encouraged to follow the example of judicial activism set by the U. S. Supreme Court, and this practice became rampant, so that now it has come to a state where right is frequently considered wrong and wrong considered right. It has carried so far that currently if one advances a just cause he now places himself in great peril if he elects to place his fate in the hands of the judicial system because he cannot depend on controlling law defining the outcome, but rather the biases and personal agendas of the judges. Our judicial rulings are in a state of shambles to a serious degree—all stemming from that initial step of ignoring controlling legal precedents to arbitrarily find "reasons" that never existed in order to obtain goals for pragmatic purposes. It calls to mind the appropriate caution, "Oh, what a tangled web we weave when first we practice to deceive."

From historical, political, religious, common sense and reason, and common law standpoints there are no bases to support the Roe v. Wade decision. The governing common law precedents are so tightly drawn there is no "wiggle room" or loopholes permitting departure from the binding legal holdings which constitute abortion a criminal offense of the highest degree at all stages of pregnancy. All rights flow from the right to life, and if a correction of our present unconstitutional stance on abortion is not reversed we will see a steady and powerful erosion of the strength of our nation. Disorders and violence are predictable consequences. We are now living in a state of silent anarchy that promises to explode as precedent after precedent is ignored, ultimately leading to a full-blown constitutional crisis.

The great danger in departing from established law in order to achieve pragmatic or ulterior purposes can have the gravest consequences. England's unconstitutional acts, pertaining to its violation of common law precedents, led to the Otis-Henry doctrines, which, in turn, stirred the American colonies to make their ominous and fateful decision, declaring "When in the Course of human events…."!!!!

CONCLUSIONS

The following observations are those certainties that are firmly supported by established law, facts, historical records, and scientific and medical knowledge:

1. The Roe v. Wade decision was void at its inception because it was crafted in contravention of controlling common law precedents.

2. In America, under current established controlling common law, it is unlawful to procure, cause, bring about, or carry out an abortion on a pregnant woman at any stage of gestation. A heavier punishment may be levied for an abortion carried out during the latter stages of pregnancy.

3. All the state laws passed in accordance with common law precedent prior to the Roe v. Wade decision and not repealed are still valid. Those states, which repealed their abortion laws, revert back to the controlling common law.

4. The right to privacy is not a factor in determining whether abortion is unlawful in America under established controlling

common law precedent. As a matter of fact, the conceptus, being a person, genetically unique and distinct from the mother, has as much right to privacy as the mother.

5. Life begins at conception, but the matter of when life begins under immutable applicable common law is not a criterion governing the unlawfulness of abortion. It is only necessary under applicable controlling common law in America that a woman be pregnant when an abortion is procured or carried out to trigger the act as a crime.

6. Under immutable established common law precedent, personhood is defined as beginning at the time of conception. Inasmuch as the United States Supreme Court declared that if personhood were established as beginning at the time of conception, their ruling in Roe v. Wade would fail, it follows, therefore, that by their own accepted criterion, the Roe v. Wade decision stands void.

7. The genetic code of the conceptus is always unique and different from that of the mother, so the conceptus is a distinct human being, and not a body part of the mother.

8. Under established controlling common law in America prevention of conception (contraception) is a crime.

9. The jurisdiction for trying abortion cases resides with the several States. No constitutional amendment, statute, or any other undertaking whatsoever can remove the controlling established immutable common law precedents that define abortion as a heinous crime. However, it is within the jurisdiction of the several States to establish appropriate punishment of the crime, sufficient to encourage its deterrence and yet severe enough to punish the actual commission of the crime. The several States may also remedy or enhance the common law governing abortions, but it is not within their power to erase the governing principle,

which declares that the abortion of a pregnant woman is a heinous crime at any stage of gestation. Appellate jurisdiction of the United States Supreme Court in abortion cases should be removed by Congress, as it is empowered to do, under Article III, Section 2 of the U. S. Constitution.

10. It is misplaced reasoning to seek a Constitutional Amendment defining personhood as beginning at conception. As stated previously, that fact has already been established by immutable common law precedent. Further, to pretend such a power to define legally when a human being is a person is open to all kinds of mischief and would mean it could be changed later at any time at the whim and fancy of the Congress or State legislatures. Not only is personhood already defined under the common law as beginning at conception, but science has also established it as occurring at that time (See Section I).

11. It is misplaced reasoning to seek a Constitutional Amendment, declaring that life begins at conception. This fact is already established by immutable common law precedent and medical science. Further, life is not something that can arbitrarily be defined as to when it begins. It begins when it begins. Under established common law, abortion is prohibited at any stage of gestation. To acknowledge (falsely) that the time that life begins can be determined by statutory or constitutional means would signify that it could be changed at any time in the future at the whim and fancy of a future generation. Life is an unalienable right that is a gift of God given at the time of conception, officially recognized by our country at its founding, and can never be lawfully taken away by any means whatsoever.

12. Under immutable controlling common law, the father has vested legal rights attached to an unborn child at all stages of gestation.

13. Disclosures made in this treatise, by their contents, may have

a chilling effect on many persons because our present culture has been conditioned to believe that no discomfiture is to be accepted which would interfere with realization of every desire without sacrifice, hardship, or inconvenience. The right of freedom has been falsely translated into license to do whatever one wishes. But, facts are facts, and those disclosed in the body of this text, pertaining to or related to the crime of abortion, are firmly and permanently fixed, and not subject to arbitrary interpretation, for example, to arrive at a pre-desired conclusion.

14. For the courts to continuously find a legal basis for abortion, in the face of the clear established common law prohibiting such practice, amounts to misbehavior by failing to enforce the law and to uphold the constitutional guarantees protecting the rights of the unborn. Unless the established law is enforced as recited in this treatise it means we will have become a nation of men instead of laws. It will mean we are in a state of anarchy which will, if not arrested now, promise to spread in scope along with its venal consequences. Ultimately, tyranny can be expected to surface because all rights find their foundation in the right to life. Preservation of the sacred right to life is the prime barrier against tyranny. If that right disappears then all other rights can be expected to eventually vanish as well.

15. It was the sacrifices our founding fathers and the early colonists were willing to endure that led to the preeminent and enviable position to which our country rose among the nations of the world. Now, to restore the health of our country and to reestablish the high moral standards it adopted through our founding fathers will require the emergence of similar strong and insightful personalities who will be dedicated to taking those actions necessary to recapture those high ideals. Because all rights find their foundation in the right to life, it is imperative to correct the present warped views on abortion that devalue life at its most fundamental level. No less than in 1776 we now need strong leaders of stature who are willing to step forward and

"with a firm reliance on the protection of divine Providence" to pledge "our Lives, our Fortunes, and our sacred Honor," if necessary, to salvage our judicial system and nation, and to restore our country to the pristine beauty it once enjoyed.

REFERENCES

1. Roe, et al. v. Wade, 410 U. S. 113, 1972.

2. Means, C. C., Jr., The Law of New York Concerning Abortion and the Status of the Foetus, 1664-1968: A Case of Cessation of Constitutionality, New York Law Forum, Vol. 14, 1968, pp. 411-515.

3. Means, C. C., Jr., The Phoenix of Abortional Freedom: Is a Penumbral or Ninth-Amendment Right About to Arise from the Nineteenth-Century Legislative Ashes of a Fourteenth-Century Common-Law Liberty? Vol. 17, 1971, pp. 335-410.

4. Blackstone W., Commentaries on the Laws of England (Chicago: The University of Chicago Press, 1765),Vol. 1, pp. 63-85.

5. Scott, M. S., Quickening in the Common Law: The Legal Precedent Roe Attempted and Failed to Use, 1 Mich. L. & Policy Rev. 199, 251-252 and fns. 147 and 148 (1996).

6. Ibid., p. 253.

7. Ibid., p. 259.

8. Coke, E., Institutes of the Laws of England (Union, N. J.: The Lawbook Exchange, 2002), Part III, p. 50.

9. Scott, M. S., Op. cit., pp. 228-229.

10. Ibid., pp. 225-226.

11. Ibid., p. 231.

12. Ibid., p. 239.

13. Blackstone, W., Op. cit., Vol. 1, p. 64.

14. www.hope.edu/academic/religion/bandstra/RTOT/CH3/HAMMURAB.HTM.

15. Ancient History Sourcebook: The Code of the Nesilim, c. 1650-1500 BCE: www.fordham.edu/halsall/ancient/1650nesilim.html.

16. Ancient History Sourcebook: The Code of the Assura, c. 1075 BCE: www.fordham.edu/halsall/ancient/1075assyriancode.html.

17. Scott, M. S., Op. cit., p. 241.

18. Ibid., p. 263.

19. Ibid., p. 264.

20. Personal communication, Dr. Kenneth Pennington, Professor of Ecclesiastical and Legal History, Catholic University of America, Washington, D. C.

21. Catholic Encyclopedia: Canon Law: wysiwyg://11/http://www.newadvent.org/cathen/09065a.htm.

22. Winroth, A., The Making of Gratian's Decretum (Cambridge, United Kingdom: Cambridge University Press, 2000), p. 9.

23. The Apostolic Constitution, Effraenatam: www.iteadjmj.com/aborto/eng-prn.html.

24. Lader, L., Abortion (New York: The Bobbs-Merrill Company, Inc., 1966), p. 79.

25. Scott, M. S., Op. cit., pp. 238-239.

26. Ibid., p. 239.

27. Ibid., p. 239-240.

28. Blackstone, W., Op. cit., Vol. 1, pp. 125-126.

29 Scott, M. S., Op. cit., p. 251-252.

30 Ibid., fn. 148.

31 Lejeune, J., Testimony in Davis v. Davis, et al., Cir. Ct. Blount Co., Tenn., Equity Div. (Div. I), No. E. 14496, 1989.

32. The Winston Simplified Dictionary, Advanced Edition (Philadelphia, 1939).

Index